PATRICK WHITE

Flaws in the Glass

A SELF-PORTRAIT

JONATHAN CAPE
THIRTY BEDFORD SQUARE LONDON

First published 1981
Copyright © 1981 by Patrick White

Jonathan Cape Ltd, 30 Bedford Square, London WC1

British Library Cataloguing in Publication Data

White, Patrick
Flaws in the glass.
1. White, Patrick – Biography
2. Authors, Australian – 20th century – Biography
I. Title
823   PR9619.3.W5Z/

ISBN 0-224-02924-X

Photoset in Great Britain by
Rowland Phototypesetting Limited,
Bury St Edmunds, Suffolk and printed by
Butler & Tanner Ltd,
Frome and London

to Manoly
again

Contents

The author and publisher are grateful to the following for permission to reproduce photographs: Brett Hilder, 1; Axel Poignant, 17; Ern Macquillan, 22; J. Wong from Colorific!, 25; Will Young, 26, 27; and Elly Polymeropoulou, 28.

Illustrations

Flaws in the Glass

It was the summer of 1926. I was fourteen, and they had taken the house at Felpham, Sussex. For my mother, the fuzz of green landscape surrounding us was English, pretty, so much more desirable than glare and drought and the threat of snakes. My father saw it as pasture for lamb and beef. For myself it meant solitude in which wounds were healed, until country voices reminded me I was a foreigner.

I was most myself in the neo-Gothic house my parents had rented for the holidays: no connection with any other part of my life, yet in it life seemed to be forming.

There was the Long Room, at one end the garden, at the other the great gilded mirror, all blotches and dimples and ripples. I fluctuated in the watery glass; according to the light I retreated into the depths of the aquarium, or trembled in the foreground like a thread of pale-green samphire. Those who thought they knew me were ignorant of the creature I scarcely knew myself.

If at school I walled myself up inside my pride, in the holidays I crept out of it. The London streets gave me confidence. My nonentity allowed me to flow with the tide. I strutted at times, enjoying anonymity amongst the pink, engrossed faces and the pallid distracted ones. I devoured the arrogance of those who had nothing to fear, insolent, tailored men, and their long lean women bleeding under cloche hats, furs thrown open on salt-cellars and meagre breasts. Their remoteness and the fact that they would have disdained me did not wither, but on the contrary, fertilised the soil in which the seeds of colonial snobbery lay.

What disturbed me without my understanding were the voices of the Welsh miners, their high harmonies as they tramped the Brompton Road in greasy raincoats. Straining towards an unattainable Jerusalem they reached me in depths I would not have known how to explore. Afterwards at the hotel I lay on my bed, tossed on a sea of Welsh voices and some unreasonable unhappiness. I got up and washed my face and hands, and went down with my family to the grillroom, to a dinner of fried whitebait in little potato baskets.

She takes me by the hand, her cotton glove. We are walking in this brown, Sydney street, between narrow houses, on hot asphalt. She is Great Aunt Grace who has come from West Maitland to stay with us.

It is stuffy and tiring.

'Come along, Patrick.' Aunt Grace is small, gentle, patient.

'Aren't I Paddy?"

'You're Paddy. But Patrick is your *real* name.'

I had seen it written on the hairbrush, but as though it belonged to somebody else. In this world of purple lantana and ox-blood brick I seemed to belong to neither of my names. Deliberately I stubbed the toes of my boots against the burning asphalt and began to sulk. I was good at that.

Not that I didn't love Aunt Grace. Her pale-brown wrinkled skin had the look of bark, but was soft to touch, like paperbark.

Great Aunt Lucy, her younger sister, I liked less. Aunt Lucy was plump and quilted. While she looked soft and comfortable, she wasn't. She breathed hard, and was inclined to snort. The Micks enraged her. She saw them as the source of most Australian evil, with the result that I took to flying past St Canice's on the corner. I never felt safe from nuns and priests, drunks, larries, or the Mad Woman, till well up the gravel drive, beyond the bunya tree.

2

You weren't supposed to believe in magic, so I forgot the Catholic witches. Then that summer at Turret House, Felpham, after a course in poisoning and general necromancy from whatever book, I moulded the wax effigy, and stuck it with pins, and threw it on a fire. I couldn't hope to know results till term began. Yet something happened which was perhaps a warning: a friend of my mother's had a slight fit in the vegetable garden where I had thrown the image on a heap of smouldering leaves. Could the spell have taken a wrong turning? It worried my conscience. I wished I could have confessed to someone. What made the situation worse, the ivy-cluttered window of my narrow room looked out on the vegetable garden, full of those melancholy stalks and fretted leaves of most rented vegetable patches. The room itself was cramped and stuffy. It smelled of saltpetre from the asthma papers I was given to burn during sleepless nights inside the ivy curtains.

The house had belonged to the poet Hayley, of no importance except as the friend of genius. He had a mad wife he used to chain to the flint columns of a summerhouse so that she could take the air. There was a cemetery with headstones carved to commemorate the lives of pets. There was a medlar tree, umbrella-shaped, under it the stench of rotting fruit.

The damp summer dragged on as I waited, half reluctant, half impatient, to discover the effect, if any, of my magic.

Felpham was a bore on most counts: the redundant children invited to stay, visiting Australians giving the details of European tours and their presentation at Court. If the adults noticed a shadowy boy, eyes set in discs of beige flannel, floating in and out of their conversation, they couldn't quite get the measure of him. Voices trying to jolly him, fell, and he would make his temporary escape, not only from the company, but his own reflexion in the glass.

I hated the appearance I had been given, but would not have known what to substitute had I been able to choose – unless something strong and handsome, as ordinary as those I despised and envied.

To fill the long stretches of the summer hols there were card games – rummy and grab; fingers trembled on the planchette as the will urged it to predict a desirable future. The best memories of those trundling weeks are of the plays the brother and sister acted, between the mirror and the garden, to an audience of amused, though proud parents, and their suffering friends. Dragged into the act by a compulsive performer, Sue may have suffered too. I remember a blank verse melodrama where, in the course of many monologues, Sir Topaze disembowelled his victim and 'flung the entrails to the winds'.

My only sister was probably the prototype of a long line of Australian Sues, healthy and noisy. Suzanne was her *real* name, with which she got stuck for more capricious reasons than Patrick got his. Our mother had been to *The Girl in the Taxi*, a musical comedy of the first decade in which the most popular number went,

> Suzanne, Suzanne,
> We love you to a man,
> We yearn for you,
> We burn for you,
> Our sweet Suzanne . . .

'Suzanne' did at least pander to the colonial nostalgia for what they believe to be French. There was no good reason for 'Patrick' beyond the fact that all the White uncles had names like Henry, Arthur, Ernest, James. Yet in labelling me, my mother seems to have disregarded the sensibility of Great Aunt Lucy with her horror of the Micks.

I don't think Sue/Suzanne gave thought enough to her name to worry. I don't think she worried, except that as a little girl she wanted badly to be a boy, and would bury her wrist with its permanent bangle deep in the pocket of her shorts. She was her school's champion bowler, whereas her shameful brother never scored more than two runs, on days when he was lucky, and always got hit by the ball if he couldn't avoid being in its way. The siblings fought bitterly till reconciled in later life by blood and childhood in common. We also enjoyed what is known as a 'good larf'.

I was approaching old age when I met the poet R. D. Fitzgerald, who recalled an incident from my childhood. His brother had married one of my distant cousins. The poet ran into the couple on a day when they were expected by my parents at our house at Rushcutters Bay. On next seeing his relations he asked how the visit had gone. 'Oh, all right . . . ' my cousin sighed, 'but that dreadful little boy was there.'

Visitors were always charmed till my sister, a pretty, dimpled child, told what I had been saying about them. I was this green, sickly boy, who saw and knew too much. If I was shy and withdrawn, it was only till provoked. Then I could answer back.

My parents were very disturbed by having a delicate son on their hands. I was kept out of draughts and protected by woollen combies. They needed me to give them a sense of their own continuity by inheriting my share in a considerable sheep-and-cattle station. A grazier's heir should have been sturdy, but nobody would insure the life of the one they had got. If I half-realised that my languor and breathlessness were causing a serious situation, I was not worried by it. All that I saw, all that was happening around me, was far too vivid for me to believe in the event which carried off old people and pets.

We cried for the cats and dogs we buried under crosses made from the spines of palm leaves, in graves stuck with wilting marigolds. The deaths of old people were casually mentioned. They hardly concerned us.

Thunderstorms were more frightening than death, and the Mad Woman, and the remark overheard in a discussion between other people's mothers ' . . . can't help feeling he's a little changeling . . . ' Nor did the laughter which followed help explain what I was or what I had done to my evidently unfortunate parents.

It was only a brief flicker of fear, like the lightning in the purple thunderstorms. There were the steamy mornings, the walk home from the baths, and watermelon at the end.

It was on the walk home from the baths, around the age of seven, that I had the first erection I remember. While looking down I suggested to my father that something unusual was happening. He became prim and embarrassed, shifted his wet bathers from one shoulder to the other, and told me to step out. At the same time there was the passing glimmer of a smile.

5

It was around this same age, after the baths, that I met my first poet, without knowing or caring much what a poet was. Sue and I had got stuck into the watermelon in what for me was the best part of the garden, all shade and spangles, under the custard apples and the guavas, outside the latticed summer-house, when our father came down the stone steps bringing some friend I had not seen before. He was dressed like any other gentleman, in a tobacco-coloured suit with a gold watch-chain strung across the waistcoat, trilby hat, and one of those stiff collars which grew grubby at the edges soon after contact with the cleanest skins. This was the driest kind of gentleman, his face like a wrinkled, sooty lemon. My father introduced his children to Mr 'Banjo' Paterson.* Whether the stranger spoke to a child whose face was buried in a slice of melon I can't remember. My father seemed proud of the acquaintanceship. I've wondered since what they talked about. They could have got together over horses, sheep, and cattle, and of course the poetry was of a kind no self-respecting White need feel ashamed of.

On those steamy Sydney mornings, my first erection and my first poet: first ripples on the tide of passion . . .

Life at Turret House, Felpham, was a succession of whirlpools: emotion, discovery, magic, the disenchantment which usually follows too feverish anticipation. Magic did not turn out to be the short cut I had hoped for. At the beginning of term the object of my curse was blander and even more vindictive while taking us for Latin Unseen, myself still the loon stooge through whom he made his points and got his laughs. I continued trudging an interminable road, or seated before a sheet of blank paper waiting for something to pour out. It never did, except in the bath, when I became conscience-ridden. What if it didn't escape down the hole and I were faced with the scorn of some knowing maid?

At Turret House, apart from my mother's faithful Mabel who followed us wherever we went, there were a rented cook, a fat, steaming, jolly young woman, and a rented butler. We had never had a butler before and Donald was not what my reading told me a butler should be. A thinnish, fair young man in an alpaca jacket, he rarely shaved, because, he said, he was a skin short and had to give his face a rest. There was the night Donald chased the cook while dinner was being dished up below. How the chase developed and how it ended remained a mystery. The children were hustled out and told to wait in the dining-room, where Mabel, red in the face, brought them their food.

The ceiling of the dining-room of the house that Hayley built at 'sweet Felpham' had been papered over with a wrinkled sky and putti seated on cumulus cloud at each corner of the firmament. One cherub had almost been hit by a rocketing cork.

I was more impressed in my childhood by Hayley of the forgettable verses, Gothick folly, and mad wife, than I was by the friend who had lived in a thatched cottage in the village, on a corner shaved by buses bound for Littlehampton or Bognor. At that stage Blake was no more than a name, if that, to the frustrated poet struggling inside me.

When I say 'poet', I did at first aspire to poetry as a means of shaping the emotional chaos with which I was possessed. Poetry was mostly what I read as a child. To an adult like my mother, who didn't read it, or had as a girl and failed to understand it, poetry would corrupt less than prose. When I say I had flipped through most of Shakespeare by the age of nine, I wasn't quite the prig it makes me sound. I understood less of the language than the average adult, but enjoyed the blood and thunder, the come-and-go, the stage directions (that magic word EXEUNT). But corrupt I became secretly on prose, on the *News of the World* and *Truth*, on *Wuthering Heights*, Ethel M. Dell and Elinor Glyn. I was found with *Lady Windermere's Fan* in my hand. The offending volume was promptly locked in the kind of glass-faced bookcase containing the books which are never read. By way of compensation I was allowed *A Crown of Wild Olive* and *Sesame and Lilies*.

Books — as time unfolded, a release from boredom through the mystery of locked words, especially at Felpham, where the weight of adolescence grew intolerably heavy in spite of solitary jaunts to hideous Bognor, the smell of low tide, the pier with its exhaustible peepshows, and touring versions of *Katya the*

Dancer, *Betty in Mayfair*, and *Saint Joan*. Bored, yet hypnotised, one lounged around, sick on milk chocolate, suddenly thudding with a sensuality aroused by pressure or glance from somebody probably unconscious of what they had provoked.

Sometimes at a distance, in a theatre or on the opposite side of the street, I might catch sight of my sister chaperoned by a female cousin or a maid. Ashamed of each other the siblings looked away.

The parents belonged to the Upper Hunter. My father was born outside Muswellbrook, my mother brought to Australia as an infant by a mother from Maitland and an English father who never settled permanently at any of the points at which he rested. Grandfather Withycombe was what the Whites must have called a 'no-hoper'; he lacked the gift for making and keeping money. According to hearsay and their photographs, my Withycombe grandparents were an imposing, handsome, violent couple; they were said to fly at people with whips. The Whites were on the whole dry, mild, close, though with a few drunks and suicides, even homosexuals, lurking in the darker branches of the tree. (One alcoholic had a habit of locking his wife and daughter in a room, then rushing demented through the house firing off pistols.)

My mother, Ruth Withycombe, and her brothers, Clem and Ralph, inherited their parents' violent tempers. I never met my Uncle Ralph. He used to throw himself on the floor and curse his siblings, with the result that they dismissed him from their lives. In their youth when Clem was running a dairy co-operative and butter factory at Muswellbrook Ralph is said to have found solace in butter-making. Years later, churning and washing the butter in the kitchen at Castle Hill, I realised I had this in common with Ralph, along with the Withycombe temper. However Ralph may have raged, Clem, Ruth, and Patrick (more Withycombe than White) must have run him close.

My father was small and mild. I can't remember him losing his temper even when forced to assert himself by taking a strap to me. The whippings were left to my mother, whose technique with a riding crop was formidable. She can't have benefited from these fruitless exercises in correction; they left her an emotional shambles. Taking into account the currency of punishment in those days, I believe I deserved what I received. Thrashings were quickly forgotten. What I could not forgive was my parents' amusement at their child's attempts to express his ideas, and their conviction that what I detested was what I would like. Even more, I resented their capacity for boring me, and my mother's relentless determination to do everything for my own good, which included dumping me in a prison of a school on the other side of the world.

Victor (Dick) White was the youngest in a family of six sons and one daughter. He was forty-two when he married; Ruth Withycombe was ten years younger. Three of his brothers had married three Ebsworth sisters from Bronte. Mum used to say Dad only married her because there wasn't another Ebsworth. I expect she was right. Imagination was not part of the White make-up.

After the wedding they spent the best part of two years travelling in Europe and the Middle East. It resulted in Paddy's reciting the European capitals at Mummy's bedside when he burst in early to disturb her sleep. There were the French irregular verbs too, but that was later.

If Dick had been round the world before without any of it rubbing off on him, Ruth was determined on worldliness. Without losing her native innocence, she did acquire quite a lot, and as she junketed up the Nile and across the Aegean, gyrating in Budapest, Vienna, Paris and London, always London, something of this determination must have transferred itself to the embryo in her restless womb.

There is a snapshot of Ruth and Dick posed on the plain of Bisley, England, in 1911. It must have been taken shortly after their son's conception. Dick is looking quite the masher, in straight-set boater and bow-tie, the Whites' blue eyes giving him that blind look. Ruth is seated on an iron chair, a thoughtful Edwardian caryatid starting to enjoy womanly fulfilment. (Apt to look coarse in the early days of her marriage, she told herself later, 'A woman must decide between face and figure' and settled for the figure.) On the plain at Bisley they were still

9

in love; they hadn't become the Ideal Couple they frayed their nerves living up to. They were still spontaneously 'Dicky' and 'Bird', unconscious, poor things, of the cuckoo they were about to hatch out.

It happened on May 28, 1912, at Wellington Court, between the stream of Knightsbridge traffic and, by comparison, the pastoral calm of Hyde Park. The White nest faced on the park. Ruth had trouble feeding Paddy in spite of the sandwiches of raw beef and celery. How they got round the trouble I don't remember hearing. There was an English nanny, but I doubt she was an expert on breast feeding, though pretty hot on protocol.

Nanny Galloway came with us to Australia when I was six months old and stayed till I was three. I remember her in a blurred way, a white, hour-glass figure divided by a buckram belt. 'Old Nanny', as she was referred to, had nursed German princelings and would describe how it was expected of her to leave the room backwards in the presence of her exalted employers. My more or less democratic Australian parents couldn't stomach it eventually. 'Old Nanny' retired to the suburb of Lilyfield, where she invested her savings in cottages, from which she wrote gushingly from time to time. My parents wanted me to visit her during my teens, but I could not face her gush. Besides, all our loyalty, our affection, was soon directed at 'Little Nursie', Lizzie Clark from Carnoustie, Scotland, who took over after Sue's birth.

My first clear memories surface in Phillip Street, in a block called 'Cromer', in those days desirably elegant, but which grew seedy over the years (semi-office semi-residential) and was finally crushed to make way for the present Wentworth Hotel. We had two adjoining flats: that in which my parents lived radiated a pink glow; its twin, for Nanny/Nursie, the child/ children and a maid was like most of Sydney at that period, predominantly brown. I seemed to drift between the pink glow from lampshades and carpet and the dull glower of brown linoleum. I sat on a chamber-pot while Old Nanny, opposite, sat on hers, encouraging Paddy to do it. I was always vomiting — liquorice all-sorts after castor oil, arrowroot biscuit over *Struwwelpeter*. I loved Alice Burgess the maid. I rode on her back as she scrubbed the floors. I was fascinated by the double-yolker she showed me in a basin as they cut my toenails on a balcony above Phillip Street. A child's wooden spade from which the digging end had broken off, she covered with silver paper and decorated with a tinsel bow. She gave it to me to use as a wand. There were puzzles: Germans living in the basement pelted with refuse by tenants higher up while chopping firewood in the yard ran back inside protecting their heads; Chidley* dressed in his white tunic looked jaunty enough as he passed along the street followed by a laughing, jeering mob. Were you supposed to laugh or cry?

Most mysterious the pink glow of the second flat in which Dicky and Bird led their other life, all cheeps and pecks. Not without its stridencies, as when Mummy caught Daddy chewing a wad of the disgusting new American gum. Daddy pretended to cry. Paddy was shocked and ashamed. Mummy seemed to rule the roost. After that they sat down to dinner, which had come up by hatch from below, while Paddy wound up his mechanical (real fur) kangaroo. It hopped across the soft pink rented carpet.

Uncle Frank of 'Saumarez' appeared at the door in a long beard. I got a fright. Uncle Clem of 'Barwon Vale' was better. Dashing. He had a brick-red skin and an eagle's beak, but a tremulous eagle. Perhaps a child made him nervous; he never had one of his own, and was most at ease in the company of racehorse-trainers, bullockies, stockmen. No one knew more about life in the country. With Sid Kirk, the sawmiller Lizzie married, he became my source of information. If each was inclined to improve on fact, their improvements were gifts to an emerging novelist in a generally unimaginative world.

Of the uncles other than Dick's partners at 'Belltrees', Frank remained a myth at Armidale. My father and I sometimes came across him surrounded by a bevy of plain daughters in one of the so-called refreshment rooms at the Sydney Agricultural Show. We seemed to make a point of scarcely recognising one another – a noticeable White trait – though Dick may have been overawed by a brother so much older than himself, who could have been my grandfather. My Uncle James I am not conscious of having met. I must have seen him. It was said he used to follow my pram in the Botanic Gardens to catch a glimpse of Dick's son. He had come to the Phillip Street flat after the return from London. Ruth offended by receiving him in a sleeveless frock; his wife Emmy, and her sisters Maud and Milly, covered themselves, as I remember, with total disregard for flesh and climate. Uncle James never relented. When Emmy died, and their daughter, another but different Ruth, invited Dick's wife to stay at 'Edinglassie' when she went to Muswellbrook for the funeral, he ignored her presence. Alas, my inability to forgive is a trait I must have inherited from my Uncle James.

Ruth said while we were driving down to Cheltenham, 'This is the proudest day of my life.' When the gates of my expensive prison closed I lost confidence in my mother, and the Uncle James in me never forgave. What my father thought I can't be sure. An amiable, accommodating husband, he went along with what she wanted.

Cheltenham was a seed sown in an ambitious colonial mother's mind by the English head of a preparatory school in Australia. Though the man turned out to be what my parents considered a 'no-hoper', the damage was done. I started serving my four-year prison sentence.

Our days and nights in the house in which I was boarded revolved round the 'sweatroom' where we did our prep and led the little social life we enjoyed in an English public school. My first impression of this sweatroom was one of varnish and carbolic, together with the smell of radiators you could press against for warmth if you were lucky enough to have one alongside your desk. One wall was panelled with lockers. Smelling of emptiness and varnish at the start of term, the lockers developed a riper, more furtive personality, which asserted itself through gusts of musty fruitcake, tantalising whiffs of orange and chocolate, and the more passive presence of damp, mutilated textbooks. Secrets you kept in your locker became open ones, you suspected; the only safe place was your head, dreams the only refuge after we were locked in the dormitory at night.

They had to give us access to a lavatory in spite of the opportunities it offered. I spent half the night in it finishing prep for which there hadn't been time in the sweatroom, tormented by the mysteries of algebra and trig, lulled by Virgil's *Eclogues* balanced on goose-pimpled thighs. Morning again: a quick slick-over, eyes and armpits, in the toshroom, before a final go at prep, and the lumpy porridge and bread-and-scrape we were served at breakfast. No wonder boys are at their cruellest at breakfast in a boarding school. Accents crop up: '. . . me dad from Bradford . . .' I was reminded of the deformity I carried round – my Australianness. I hardly dared open my mouth for fear of the toads which might tumble out, and the curled lips, cold eyes waiting to receive renewed evidence of what made me unacceptable to the British ruling class.

Any boy who had not thought about sex must soon have been made aware of it, such was the housemaster's obsession. Perhaps understandably. Shortly before my arrival the poor wretch had weathered a scandal when half his house had been expelled. He would burst into toshroom or gym hoping to catch us *in flagrante*. He was the tallest man I had seen. He smashed the light bulbs caning us. He promised to stamp out a 'morbid kink' on discovering my passion for Chekov, Ibsen, and Strindberg, and only stamped it deeper in. Never during my stay in his house did he uncover sex, though he must have disturbed fantasies in his forays through toshroom steam and the stench of sweat-sodden jerseys and mud-caked boots in a more puritanical gym. We were far too frightened, I think, and at least one of us found the climate uncongenial. Even in the more brazen days of my maturity, English sex shivered and plopped remorsefully like a gas fire on its way out. The strength of the game lay in the opening gambits.

When they considered I had settled down at my public school my family returned to Australia. The parting took place on a Swiss railway platform. At one point my mother withdrew with my sister, leaving my father to acquaint me, I began to sense, with the facts of life. Dusk was tingling with village lights and their refractions off snow as we plodded up and down in our felt boots, wordless banners of white mist ballooning out of our mouths. My heart was beating horribly, but at least it wasn't up to me to speak; the onus was on my unfortunate father. At last he accepted his duty. He warned me against the seats of public lavatories. We were both breathless with relief at the removal of a difficult situation. Only the light was nagging at me, the sounds of a train preparing to burrow through a Swiss landscape towards bleached grass, eroded creeks, and the wounding blows from butcherbirds' beaks. The wounds I suffered on the snow-bound platform were of a duller kind which promised suppuration. I was determined to keep my grief within the bounds of that manliness I was being taught to respect, when I would have liked to tear off the rabbitskin glove he was wearing and hold the sunburnt hand to my cheek. I did nothing. I didn't cry. I only throbbed as a windowful of faces slid away through the Swiss dark.

Kind friends escorted me back to resume my sentence at Cheltenham.

Would I have felt sentenced in Australia? The masochist in me might have seen to that. As it was, memory helped flesh out an English schoolboy's idyll: riding a pony bareback through girth-high tussock, stripping leeches from my body after a swim in a muddy creek, my solitary mooning through a forest of dripping sassafras towards the sound of the waterfall. My parents played no active part in this country of the mind. I clung to them as a lifeline. I wrote them each week a stilted childhood letter. I was a dutiful son, if not the kind they would have chosen, some doctor-grazier, cricketer-barrister, or my mother's version of a diplomat; we were all three guilty and innocent parties, suffering one of those betrayals by fate.

I wrote to 'Nursie' too. All the genuine love in me was directed at this substitute for a mother. Lizzie Clark came to us shortly after Sue's birth, when I was three. I hated her at first because she brought change. What with the baby and the new nurse it was a time of upheaval. I was hateful and destructive. I stamped on her toothpaste as she unpacked. I let the water out of the hotplate as we sat at our dinner of boiled brains. Nanny had always fed the animals on the mantelpiece before I would accept a mouthful. Nursie refused to take on the animals. She even made me feed myself. I suspected her dark face, the curved nose with a distinct hole or single pockmark on one side, the glossy black hair, the Scots accent. I don't know how she won me over, perhaps by the wet kisses she planted on my unyielding mouth, drawing me out of myself until we were united in a common wet.

I loved her. We both did. We nearly cried our heads off when, on one of her afternoons off, silly Eliza told us Lizzie had gone for ever. Lizzie never had a child of her own, except a stillborn one late in life. She considered we were hers, as in fact we were.

Ruth, the titular mother, bound us to her by a series of surprises such as changes of dress, presents, tantrums, powers of organisation, and bursts of general knowledge. But she wasn't always there, what with the committees, the fittings, the luncheon and dinner parties, or else she was lying down with a headache or falling arches. Except when a little child, I don't think I loved, I only admired her after a fashion, until I pitied an old bedridden, half-blind senile woman, and pity is a pinchbeck substitute for love.

As for Dad, 'Bird's Dicky', he soured with age. From being that bland, equable fellow with the blind blue eyes and dossy bow-tie, he developed a telling line in sarcasm. During the middle part of their lives, 'Bird' and 'Dicky', the Ideal Couple, sat side by side at their desks in the smoking room at 'Lulworth', discussing investments, children, servants, the shocking behaviour of Jack Lang,* while Ruth answered the invitations in her full, assured, feminine hand, and Dick put his angular signature to the cheques, or wrote to his brothers Henry and Arthur in the inhibited script of the early Australian male. Ruth and Dick were seldom apart. It looked a stuffy relationship, till I realised later in life that it was one I coveted.

I might have loved Dick had I dared, and had we been able to talk to each other. Did Ruth ever guess? Years after both were dead, Lizzie was left practically a mummy, anchored by arthritis, glaucoma, and senility, to a chair in a home at Wentworth

Falls, where anyone from the past might appear: 'My father's been here this morning; he was standing over there in the corner,' or 'Walter and Robert took me out in the boat to watch the herrings,' or 'Sid's up mending the roof. He'll be in soon — he wants to see Paddy.' More often than not I didn't exist for her in the disguise maturity had given me. On one of these occasions Lizzie came up with, 'Mum was good when she was good. She was different after you had left home. The trouble was, you both wanted to be the star.' It was soon time for me to go. We exchanged one of the wet kisses practised in my child-hood, and I ran down the hill and bought a meat pie, and jumped on the Sydney train, scalding my mouth on hot gravy and remorse.

Till well into my life, houses, places, landscape meant more to me than people. I was more a cat than a dog. It was landscape which made me long to return to Australia while at school in England. It was landscape more than anything which drew me back when Hitler's War was over. As a child at Mount Wilson and Rushcutters Bay, relationships with even cherished friends were inclined to come apart when I was faced with sharing surroundings associated with my own private mysteries, some corner where moss-upholstered steps swept down beside the monstera deliciosa, a rich mattress of slater-infested humus under the custard apples, or gullies crackling with smoky silence, rocks threatening to explode, pools so cold that the breath was cut off inside your ribs as you hung suspended like the corpse of a pale frog.

On the mountain there were always threats of explosion, whether natural or human: a telephone torn from the wall by lightning, the eruption of a bush fire with scrolls of smoke and pennants of dirty flame advancing through shaggy scrub. I often flung stones at human beings I felt were invading my spiritual territory. Once I set fire to a gunyah to show that it couldn't be shared with strangers. Years later I persuaded myself that I hadn't been acting merely as a selfish child, but that an avatar of those from whom the land had been taken had invested one of the unwanted whites.

My decent parents were disgusted, never more than when I

ran away and hid in the bush to avoid a Christmas service held in the asbestos church. Ruth despised the Mount Wilson church for aesthetic reasons: 'I'd burn it down if it would burn.' If she were alive today she might have got some satisfaction from reading that we are being poisoned by asbestos. But on Christmas Day 1923 the asbestos church, erect amongst the tree ferns she rooted out methodically in establishing her English garden, was accepted as a spiritual ally in accusing her infernal child.

Boarding school, I was told, would put me in my place. Even Lizzie was indoctrinated, though I couldn't believe she really believed. The school to which they sent me was far enough from Sydney to foment terror in the heart of a timid introspective child, anyway in the beginning. It was patronised by the grazier class and Sydney families with social pretensions. The climate was bracing. The boys, like those of any school, were said to love everything about it. The building was a mansion in the Thames Valley Tudor style favoured by the Australian rich in the earlier part of the century. It had been an impressive 'residence' when lived in by an important family. By the time I arrived it was looking somewhat abraded, standing in an arid garden in which only the fittest had survived. Yet I developed an affection for its tunnels of droughty laurels and tortured golden cypresses the closer I became acquainted with them. On one side there was a hedge of blue-black *pinus insignis* which behaved in accordance with Gothick tradition, tossing and soughing when a wind blew, radiating cool at the height of a blazing summer, daunting chilblained flesh in the depths of callous winter, its own limbs bleeding at any season wherever wounded. I can still visualise the blood of those pine trees congealing in silver-grey gouts where initials had been carved or schoolboys' vicious boots had scarred them. In the Tudor mansion, between the Gothick pines and windswept, sun-bleached paddocks, I suffered my own first wounds and began unconsciously to learn about the politics of life and love. As brutes in knickerbockers hurtled past me the first night, I planned to sell my pencils and buy a train ticket to Sydney. Somehow I survived and became a different kind of brute, far sharper if I had known.

One of the frustrating aspects of childhood is not being able to peel off the webs of mystery which cling to certain events and your own haphazard presence in them. The headmaster had a fiery face and a globular figure. I can't remember his appearing in the classroom, but he supervised our cricket games, sometimes taking part in them, when he thwacked the ball so hard that even I was impressed. When not performing, he sat among the boys in his white flannels, glaring at the game, smelling of men. One alarming evening he stood in the centre of his study, fiercer of face, the smell of a man more pronounced as he held me against his stomach instead of using the cane a junior master had recommended. When I realised I was forgiven I dissolved in a mingling of gratitude and anti-climax. He disappeared from our lives soon after, without adequate explanation. He was ill, I was told by the matron through a welter of adenoidal sighs. Would he come back when he was better? No, he wouldn't. From an early age I was inquisitive: anxious to hear, to see, to open any unopened door, to experience, to know. So I was aggravated now by not knowing. Though of course in a sense we all did, and we expressed our scorn as tribal convention demanded. His dismissal had been discreet enough; our schoolboy ribaldry surrounded his 'illness' with a ridicule to drive him far beyond the pale. Publicly I went along with the tribe, but remembering the rumblings of a warm stomach, my private mind accepted him as a sympathetic character.

The next head was a disaster of another kind. He was wet, English, and had ideas about education. He tried appealing to the reason he presumed lay buried in us, but stirred up the fiends of unreason instead. The peak of his folly was exposed when he introduced phonetics into the pronunciation of French. French was sissy; who wanted French anyway? Undeterred, he distributed pocket mirrors to a class of Australian boys. We were encouraged to watch our own mouths while pronouncing aloud the despised French vowels. I who had learnt enough French from a Swiss protégée of my mother's felt too ashamed at first to join the knockers, but finally could not resist the game of destroying a civilised language and the man who dared submit us to it. We sat with our mirrors, hooting, gibbering, jeering at our own distorted mouths while the man who had triggered us off sweated at every pallid pore, a tremulous smile hoping to turn his gaffe into some kind of feeble joke which might appease his persecutors. Whoever may have experienced guilt, we wallowed at the time in our own superiority.

Less theatrical, though probably as humiliating, was the night when pupils and local bigwigs gathered in the main classroom to listen to the crystal set the head had bought as a surprise for the school. As we sat prepared for initiation into the mystery of the cat's whisker, experts twiddled and twiddled – but nothing happened beyond a spasmodic yawning of static. Afterwards at least we tucked into a supper of sandwiches, meringues, and sponge.

His only successful stroke was the appearance one morning of Madame Melba, no less, in the same big classroom. We clattered to our feet as she paraded in, dressed in brown, a springy queen of toads hung with diamonds at 11 a.m.

Coming at once to the point, she commanded, 'Hands up everyone born in Australia?' No nonsense about Nellie.

A matter of shame to a small boy, I couldn't hold up my hand with the others.

'Where were *you* born?' She pointed sternly at the outsider in the front row.

When I told her, she muttered with the knowing glint of the professional expatriate, 'Not a bad place either.'

She did not let us hear the famous voice, its pure lark-notes known to me from the cracked gramophone record at 'Lulworth'. Instead she gave us a half-holiday, which probably endeared her more to a rabble of schoolboys than any rendering of *Home Sweet Home*.

Those who are doomed to become artists are seldom blessed with equanimity. They are tossed to drunken heights, only to be brought down into a sludge of headachy despair; their arrogance gives way to humiliation at the next curve of the switchback. This applies particularly to artists of the theatre. Most children have theatre in them. Those who carry it over into adolescence and, more or less, maturity, commit the ultimate indecency of becoming professional actors. If I didn't go all the way, I became instead that far more indecent hybrid, a frustrated one. Sexual ambivalence helped drive me in on myself. Lacking flamboyance, cursed with reserve, I chose fiction, or more likely it was chosen for me, as the means of introducing to a disbelieving audience the cast of contradictory characters of which I am composed.

Theatre and magic, whether my misfired attempt at finishing off the master who took us for Latin Unseen by sticking pins in a wax image, or the vision of the child-fairy emerging from an enormous panto rose to swing above our heads on a wire in what used to be known as the Grand Opera House near Central Station, made me vaguely conscious that I was in some way involved with a world of illusion, half terror half delight, like those orgasms in a hot bath and the near ejaculations of fear which accompanied my encounters with the Mad Woman.

The Mad Woman was real enough because I first came across her by broad daylight in our back yard rootling through the garbage bins. But even by daylight she seemed to belong to some nether world rather than the realities of poverty and hunger. I could connect her blotched, alcoholic skin and munching gums with the frosted glass of pubs I hurried past at the Cross, but the stinking fish skeletons and heads she was sorting and wrapping in greasy paper had undergone some magic change. I would recognise them years later as paraphernalia of the illusion referred to as art.

By day, above the garbage bins, the great hat she always wore looked insignificant, extinct. Dusk was when it flowered, becoming for me the distinctive symbol of the Mad Woman, its huge targe apparently constructed out of the trumpets of grubby, wrinkled arums. Although she appeared to me in dreams and waking fantasies I can't have met the Mad Woman more than three times in the flesh. There was the occasion at the garbage bins when the fish skeletons and heads dangling from her fingers took on a significance above daylight and reality. There was the evening when walking with my dog through my private territory the lower garden, in a gloom of hydrangeas, guavas and custard apples, she loomed at us, swaying in a stately manner I didn't associate with drink while knowing from what I had observed that she must be drunk. She gave us a green, dreamy smile. The strange thing was she took it for granted that this part of the garden, which I considered mine and resented other children entering, was hers as well, perhaps even more essentially hers. The terrier seemed to accept it. He didn't bark or attempt to attack. When we went up to the house to tell them the Mad Woman was down in the garden, then he began to bark and skirmish. I stayed behind when he ran down with them to chase the intruder, but by then she had disappeared, except from my mind.

The third occasion, more frightening because more violent, was also at dusk. It was during the season of guelder roses, and I came across her tearing them by handfuls, stuffing them into a carrier bag, their bracts fluttering and falling amongst the privet. My mother was not far behind me. She ordered the Mad Woman to stop destroying her guelder roses. She was not obeyed of course. So she called for Solomon Rakooka, the Solomon Islander who had worked for the Withycombes at 'Piercefield' while 'Miss Root' was a girl. Sol came down the drive looking very black in the dusk, laughing for this big joke, squelching in his badly fitting boots (he never walked happily in boots; he had bunions and his arches were gone.) Sol grappled the Mad Woman in the shadow of the guelder roses, just beyond the bunya-bunya where, at another time, Mrs Bonner ordered the carriage to stop and they picked up Voss. Now Sol and the Mad Woman were wrestling and spinning in the dusk, she hissing, Sol shouting through his ragged moustache and brown stumps of teeth. Then the Mad Woman's skirt fell off. I did not see what happened after that. I ran away. I ran upstairs. I lay on my bed. The glass above the dressing-table showed me palpitating in green waves. My heart was beating, a wooden, irregular time, as in another situation the hooves of Voss's cavalcade drummed their way down the stairs in this same house.

They came and told me Sol had sent the old thing packing and I must get myself ready because we were going to the theatre. It was a popular musical comedy. As the heroine sang about the silver linings she saw while washing up the dishes I could only see the Mad Woman, the snowy bracts of guelder roses falling in a storm as she wrestled with Sol. All round me in the plush tiers of the theatre, families were offering one another chocolates and smiles and enjoying the predicaments and final metamorphosis of a waif-heroine into the wife of a millionaire. For the first time I was a skeleton at the Australian feast. I could not have told about it, and went out of my way to present the normality and smiles expected of me, while drawn back into the dusk, and storm of shattered guelder roses enveloping the Mad Woman and myself.

I loved Sol, the agent if only temporarily of the world of light and right, who had routed the Mad Woman along with my still incalculable shadow. Sol was an ex-seaman. How he came to berth at 'Piercefield', the Eden of my mother's recall, on the Denman Road outside Muswellbrook, I can't remember. He was a detail in a tapestry she would drag out whenever she needed to hold her own against the Whites. At 'Lulworth' Sol continued addressing my mother as 'Miss Root', a habit which never failed to delight us children. Ruth told how she used to call, 'Sol, why are the dogs barking? Is somebody coming?' (I believe my mother shared my hope that the course of life would be changed if only somebody would come.) But Sol would call back, 'No, Miss Root, they're barkin' at the paddymelons.'*

I grew to love Sol for mystical as well as materialistic reasons. He used to fetch me from the kindergarten where I had started spending my mornings. No other child could boast of a black attendant, and more important, on the way home, Sol would buy me forbidden sweets, garish in colour, baroque in form, and guaranteed to rot the teeth. Neither of us ever mentioned the sweets to those who might disapprove. It was our tangible secret, and on a morning when the ritual of the sweets was interrupted by my mother's arriving instead of Sol to fetch me from school, I spat in her face in Ocean Street. On reaching home I got the riding-crop, the bone handle, and was not let appear as one of the blackbirds baked in a pie at the school breaking-up. Instead I received a visit from Alice Burgess our former maid, now married, who was wearing a black velvet hat on which a sparkling butterfly had alighted. She allowed me to touch her butterfly.

The more important, the mystical tie with Solomon Rakooka, arose from his having sailed the seas and called at foreign ports, where he had collected a variety of objects kept in a box in the small brown room opening on our back yard. I never cared to go inside the room with its rank blankets on an unmade bed, but Sol used to bring out his box and we sat together on the step going through his treasures: shells I remember, a tobacco pouch with twist top and the head and torso of an Indian as decoration; there was also a vicious-looking knife from Buenos Aires. I don't think anything much was ever said by either of us. Sol would be smoking his pipe, while I dreamed and wheezed, on the precipice above Roslyn Gardens, long before they rubbished the houses below and stuffed the space with blocks of units and motels in lavatory brick.

Sol had to leave eventually. He was a drunk. His drinking bouts used to land him in gaol, and sometimes he brought home sleazy women of the type I had seen hanging round the pubs at the Cross. No doubt his own drunkenness had persuaded him to see the Mad Woman's intrusion as a joke rather than an outrage to Miss Root's possessions and respectability. When he left, we all cried. Even Dad put on a performance, as he did when we buried pets, and as he had when Bird turned on Dicky for coming home chewing that disgusting American gum.

None of the gardeners who followed Sol had his panache. Some of them were drunks too, but burnt-out Irish ones, others of an impenetrable Anglo-Saxon pseudo-sobriety.

It was a long time before I was conscious of connecting boredom with undiluted Anglo-Saxon blood. The Whites and the Withycombes were English enough, Somerset farmers on both sides, my father and mother second-cousins. The only forebear who titillated my imagination was a Withycombe said to have been fool to Edward the Second. Thought of him set my own little cap and bells quivering with pleasure. There was also a vague legend which linked the Lipscombs, Ruth's mother's family, with the Pitts. That left me colder till I heard about Lady Hester Stanhope.

Lizzie our nurse taught us that the Scots were above the

English, and because we loved her dearly we believed her. Her own dark complexion, beaky nose, and glossy black hair our mother told us probably came from the Spanish Armada wrecked on the Scottish coast, and Lizzie was pleased enough not to protest.

Uncle Clem Withycombe, for whom I nervously cherished an undeclared affection, was another beak, with dark, flashing complexion, in spite of the Somerset yeomen, and Sid Kirk, Lizzie's sawmiller husband, who taught me all there was to know about the gullies with which Mount Wilson's flanks are gashed, was a quarter French behind his drawled Australian voice and Digger's face.

Then there were the Germans, in my early youth discreet about their origins, only plucking up courage in their choice of names. There were the Morrices, my godmother Gertrude, her younger sister Minna, and the old mother with whom they lived. The boys all had the most unequivocal German names and lived in far places like Queensland. I never met the boys. There was a third sister, Lottie ('Lottie sleeps with her eyes open') married to Etienne de Mestre the horse-trainer brother of Roy the painter. I first met Roy in London in the 'Thirties, when he became my friend and one of the more important influences in my life. From him I learnt most of what I know about painting.

Old Mrs Morrice was small and self-contained, with wrinkles and rings, and compact hair like steel wool. She seemed to spend her life sitting on the edge of a hard sofa, or at any rate that is what she was doing whenever I visited them after a long hot walk across Rushcutters Bay Park and up Loftus Street, Darling Point. The brown summer heat and the cracks in Sydney asphalt became part of me, to last for ever. But it was refreshing to arrive and sit on the edge of the hard sofa with my hostess, and sip a glass of cold lemonade. The great thing in Mrs Morrice's favour was that she treated a child as her intellectual equal. We discussed Shakespeare, we read *Hamlet* together. She spoke the most precise English, with that curious, rustling 'r' which also distinguished her daughters' speech.

Minna was golden-skinned and given to ripples of laughter such as I haven't heard from anybody else. Everybody liked her, whereas people had reservations about the more withdrawn Gertrude. Gertrude varied in appearance. She was another of the dark ones. I can remember her at her best under a pile of black hair, in a white broderie anglaise dress, creamy eyelids, too white a face streakily powdered, smoking with fastidious intensity the cigarette she was holding in a little pair of golden tongs. At her worst she was wooden, oblong, in a striped voile dress, like a deck-chair come to life. Bobbed early and irrevocably, she may have been in love with Clem at one stage. She remained a spinster, one of Ruth's acolytes, helping out with the sewing. She must often had felt humiliated, but never let it be seen. She took her godmotherhood seriously, though I don't think she had a religious faith. She introduced me, book by book, at birthdays and Christmas, to Aldous Huxley and D. H. Lawrence, starting me off on whatever intellectual life I have had. At the end I used to visit her in a so-called convalescent home where I sometimes succeeded in turning on one of those dusty, obsolete gramophone records: 'Yes, it's a fine day . . . a fine day . . . a fine day . . . ' or, 'Minna's over at the barn . . . Minna came . . . Jack is with her . . . Jack . . . at the barn . . . ' Jack was a kangaroo dog I used to play with when the Morrice girls and their parents were living at 'Browlie', a low-built, biscuit-coloured house at Moss Vale, of the type in which the not-so-successful lived in the early part of the century. 'Browlie' was far too shady and probably damp. We loved the little cornflour cakes Auntie G., who hated cooking, made specially for our visits. I loved the smiling, wedge-headed Jack, until here was Gertrude, this old cracked gramophone record, the needle constantly sticking in the groove, 'Minna's here . . . the barn . . . the barn . . . Minna's come . . . and Jack . . . Jack . . . ' Sometimes I think the difference between old age and youth is that in the one a person has come back and in the other some hoped-for being is still expected.

In her old age, in the dormitory at the seedy home, Gertrude Morrice had developed a doubting smile. The passive wrists lying on the chenille bedspread had grown wisps of long black silky hair. My godmother was the starting point for Theodora of *The Aunt's Story*, which I like to see as some return for her unobtrusive opening of windows in my often desperate youthful mind.

Reading, writing, always reading and writing . . . Round the age of nine I wrote a poem called *The Tramp* which expressed priggish sentiments. At least they were not so much my own as those expected of me. When Sue produced a poem so much subtler than my suddenly pathetic verses I was duly shattered – till discovering that she had copied it out of an anthology. At that age I wouldn't have expected much of a sister, but what took the edge off my scorn was the suspicion that Lizzie, that paragon of Scottish rectitude, had encouraged her in her deceit. Lizzie may have seen it as a chance to deflate false pride, for her favourite slogan was, Never blow your own trumpet, a warning which has echoed through my life to the present day, when trumpet-blowing has become one of our favourite national pastimes.

I may have learnt some kind of moral lesson, but nothing could stop me writing. At school I helped edit, and contributed serials to half-a-dozen handwritten magazines. I began writing plays: a melodrama *The Mexican Bandits* in which practically the whole cast was shot dead in a final court scene; *Love's Awakening*, a domestic drama where a husband went out to buy a divorce, had supper with the 'other woman', and decided he preferred his wife; a tragedy in verse about a Florentine tyrant (female) who kept a cellarful of lovers. Weekly visits to the pictures at the old King's Cross Theatre, together with my indiscriminate reading, novels like *The Forest Lovers* and *The Hundredth Chance*, both found surprisingly on the shelves of the school library, copies of *Truth* bought secretly over the garden fence, and the *News of the World* borrowed from obliging English servants, all helped influence these dramatised novelettes.

I wrote a prayer, too. My father found me intoning it in the summerhouse in the lower garden. Dick in whom I don't remember catching a glimmer of religious faith though he lowered his voice sightseeing in European cathedrals, and came to church with us at Christmas to please Ruth and set the children an example, was shocked by my rites in the summer-house. In the green gloom, amongst the moss-encrusted wood-work, my behaviour must have appeared semi-pagan, almost diabolical, to a decent extrovert Australian father.

At school we were walked across the paddocks to the Bong

Bong church, nesting maggies and butcherbirds diving at our boaters. The service, honest-to-God C. of E., was taken by a hearty clergyman in Nietzschean moustache beneath a slightly popish blue ribband affirming that 'God is Love'. The pews on the opposite side of the church were filled with girls from a neighbouring school with whom we exchanged giggles and blushes. Afterwards, under the pines, standing about in awkward groups on a mattress of dead needles, sisters and brothers from the two schools made inhibited conversation before starting the walk in opposite directions.

Less formal were the scripture lessons taken at school by a retired rector who arrived in his pony-drawn sulky wearing a shovel hat. While the old man droned on, oblivious to anything but his subject, his pupils were sticking compasses into their neighbours' thighs, masturbating one another under cover of the desks, or asking permission to 'leave the room', actually to find out whether one of the pregnant guinea-pigs had given birth.

If 'God is Love' was the slogan of the Bong Bong church, Sex was the theme developed in the dormitories, in the tunnels of drought-stricken laurels, and the long grass hedged in by hawthorn. Often barely explicit, like a crush on the music mistress as she smelled a bunch of violets or guided one's hands at the piano, or spasms of admiration for a sportsmaster's hairy, muscular arms, there were also brutal, boyish orgasms. I imagined I was in love. I suffered my first agonies of sexual jealousy.

Looking back from the point I have reached, when the incidents themselves have become blurred, I can still experience my schooldays in the country as a synthesis of living sensuality: the blaze of light, a horse galloped through flickering leaves, callous flesh streaked with mud, tasselled with leeches, and the smells, especially those of crushed ants, smoke rising from twigs and bark kindled in the open, bread and mushrooms frying in biscuit tins on a schoolroom stove, hot darkness and spilt semen; and through it all I hear the birdsong, alternately cynical and sweet, piano notes faltering on frosty mornings, treble laughter intertwined with the harsh braying of broken voices.

A murder was committed the other side of the wall of *pinus insignis*. He rode into town with a prescription; his wife was sick. He bought her a bottle of eau-de-Cologne, drank the eaudy on the way home, then shot her, and a few of those who came out to take him in charge. We were not supposed to know. I experienced it along with everything else.

Sue and I had in many ways a protected childhood. We were Dick and Ruth's 'pretty baa-lambs'. Sue was less affected by it. In spite of throwing the odd tantrum when made to wear wool (it pricked her) she was more equable than I. People liked her. Friends at school lasted a lifetime. She was what I learned they call 'normal'. I could not have shared a secret with her. For that matter, I didn't share secrets with anybody. Some of them festered inside me during the asthmatic summers and bronchial winters. When she was older the normal Sue developed asthma; perhaps she had her secrets too, somewhere in the wheezy depths.

As children we fought and scowled at each other. From being a pretty, dimpled little girl she went through a hobbledehoy phase. I resented my sister's exaggerated larrikin attitudes. She had contempt for what she must have seen as my prissiness and secrecy. She tore up my books. I half-throttled her once and received the conventional punishment. Next day the swellings on her throat were diagnosed as mumps. Presently Ruth was mumping too, grey-skinned and bleary in her Antiphlogistin poultices. Contrition set in and I was sent off to cousins at Mount Wilson.

Bliss at last. The cousins had been planning to motor to the Riverina, so I was left to the married couple. Through the Davieses, Matt and Flo, I came in contact with real life as opposed to the counterfeit offered children by most grown-ups of our class. Every moment in their company was vital: the jokes, the laughter, the anecdotes, the smell of Matt's little black pipe, the scent of Flo's scrupulously laundered aprons and dresses, tattered copies of the *News of the World*, the dropscones she taught me to make kneeling together at the hearth.

Matt came from Horton-in-Ribblesdale where Yorks becomes

Lancs. Flo was a girl from Woodstock outside Oxford. Matt had been a footman at Blenheim Palace. Both had worked in Park Lane, he as footman, she as housemaid at the old Dorchester House. The footman became our cousin Owen Wynne's batman in World War I. Matt married Flo at the peace and they followed Owen to Australia. They showed no signs of suffering by their transplantation. Their lovely black sense of humour must have seen them through. They stuck it out till death in their eighties, still full of what I see as the sap of life in spite of great physical pain.

Matt had a long comedian's face. He specialised in music-hall bawdy, of which Flo pretended not to approve. I can still hear her, 'Ooh, *Matt!*' if she felt he had gone too far, while enjoying it as much as I did. We were united in our approval of Matt.

Matt could do almost anything, it seemed: kill and dress a pig, cut out loose covers for the local chairs and sofas, wait at table, tend an orchard. Flo was always kneading something, sleeves rolled back from plump white arms, a dash of flour in her apple-blossom cheeks. Flo was still kneading in her eighties, standing at her kitchen table in spite of a leg no longer recognisable as such, still able to find life hilarious. Matt was more subdued by then, though active after a fashion, dragging himself by grace of aluminium supports amongst giant azaleas and rhododendrons, and columbines a deeper blue than I've come across in any other garden.

As I aged, my visits to what had been the paradise of my childhood and youth became waking nightmare peopled with familiar figures deformed by time and arthritis. Everybody creaked. I began to hear a clicking in my own bones. The sassafras I considered mine seemed to reject my intrusions; I was surrounded by a quizzical silence, watched by invisible birds. I did not attempt to re-enter the smoky gullies I had explored with Sid Kirk as guide. What I sensed was not so much hostility as the indifference of a timeless landscape to human limitations.

All the real inhabitants of Mount Wilson as opposed to the fashionable rich who came and lived in modernised houses and left on exhausting a whim, had absorbed, I felt, something of the mountain's fatality. The spirit of place strengthened them as they dragged out their increasingly painful physical existence. They knew they belonged there.

Apart from the excursion to France and Belgium in World War I, Sid Kirk never left the mountain. Then, in his eighties, they put him in hospital at Katoomba with a heavy cold which could have turned to pneumonia. He was dead two hours after admission.

Lizzie, after leaving the Whites in 1925 and marrying Sid, kept to the mountain, minding her own business, for the best part of forty-five years. Her passion for discretion (I've noticed it in others of Scottish blood) left her practically friendless. I think she stopped regretting her stillborn daughter. Sue and I were her real family, as, in my case, she was my real mother. She loved her garden too. Most of her day was spent in it after she had read the *Herald*, every word of it, particularly the deaths. Part of the afternoon was taken up with locating the nests of truant hens. She complained that her hens were tearing up the paths and eating the concrete. Another complaint was that neighbours accused her of sitting at home doing nothing; when there wasn't a speck of dust in her house. Cooking was what she hated. Her speciality was boiled mince, with boiled beetroot if you were lucky and Sid made her bring it out of the fridge. After she had grown vague, and the dust gathered in her speckless house, the little heaps she swept together and forgot about, they dug her up and re-planted her at Wentworth Falls, where her sight deteriorated, and her mind. She died at the age of ninety-six, still part of her garden at Mount Wilson, when not watching the herring shoals at Carnoustie, Scotland, or taking us to gather dolichos seed in her parents' back yard at Rose Bay, Sydney. They brought her home to the mountain. We buried her in its purple soil, amongst the tree ferns, outside the asbestos church.

Mariamne Wynne, our cousin Owen's English wife, was another who developed a close relationship with the mountain, but a cynical one, against her will. Strolling amongst the deodars and

cedars, the eternal azaleas and rhododendrons of her labyrinthine garden, she thrummed with ironic indignation on
mentioning 'the ubiquitous *gumm*', yet in her old age I sometimes caught her, eyes bleary with glaucoma, staring at the
smoky distances as though united with them.

When she arrived in Australia after World War I Mariamne
was a dashing figure – gay in the word's archaic sense. She was
brought to 'Lulworth' to face relations. Her style, as I remember, was that of the English upper class slattern who
doesn't give a damn for anybody. She had driven an ambulance
during the war. She was dressed in black though practically a
bride. At our first meeting she was wearing monkey fur. It
straggled down from the collar of her coat like tendrils in a rain
forest and tumbled from the brim of her cloche in imitation of
the rather untidy human hair underneath. Her strong suit was
irony, which later life congealed into scorn. But to my child's
sensibility the English bride was a dazzling creature.

Though Australian bred, Owen had grown up in England,
where his mother had fled to escape from a hare-brained
marriage. The son was the picture of an English officer and
gentleman: clipped moustache, ruddy complexion, hair crinkly
as an Airedale's coat. Some found him dull, but physically
forgiveable. He became the staid, honourable husband, of cor-
rugated brow and few words, happiest in his workshop, where
he made from his own timber the blackwood furniture for their
house and panelling for the dining room. He was a professional
as a cabinet-maker, though never more than a dilettante farmer;
while his wife, the dazzling amateur, threw off performances
after dinner, impersonations of London actresses her one-child
audience had only read and heard about. She helped me make a
toy theatre, with cut-outs from the *Tatler* and *Sketch* as actors
and dancers. Lopokova was one of my stars.

Years later, faced with my writing the kind of novels she had
not been brought up on, Mariamne saw me as a target for her
disapproval along with the 'ubiquitous *gumm*'. After reading
The Tree of Man she crossed the mountain on foot to give Lizzie
her opinion. 'Does he think he's pulling our legs?' Lizzie murmured something like, 'It takes all kinds . . . ', but when the
visit was repeated after *Riders in the Chariot*, found courage to

reply, 'Some people haven't the education,' and the sister of a British ambassador retired offended, perhaps even defeated.

Mariamne was one of those mothers who dream of their daughters marrying, if not a lord, at least one of those Honourables who used to put in time as A.D.C. to a Governor. When the girl does better for herself by accepting, shall we say, a nurseryman, the mother broods and festers. But she had her forgiving moments in her painful, solitary last years. The chatelaine would drag herself out to where the terrace and its frayed chairs faced the never-ending vista of gums, and laugh and joke in her close-lipped, grudging, English fashion. Dressed in shaggy grey woollies she would shake herself like an aged dog cheered by a visitor's attentions, then as the inevitable parting approached, lapse into sad-eyed apprehension that those she thought she could trust might be abandoning her for ever.

In spite of the ambulance, the monkey fur, the impersonations of actresses, Mariamne was an English puritan, as Lizzie was a Scottish version of the breed. In the circumstances Matt and Flo, with their vein of Elizabethan bawdy and echoes from the late Victorian music-hall, were an exhilarating antidote. All four of them were exotics in Australia, adaptable transplants in some cases, at least one reluctant to submit to rude conditions in antipodal surroundings. Each is a fragment of my own fragmented character. And Sid Kirk, the ex-Digger of French extraction, showed me the lyre birds, the wombat tracks, zircons in the trickle of a creek; he taught me to unravel bush silence. Somewhat surprisingly Sid took a dry interest in local scandal at the same time. He stimulated my novelist's imagination in a way that Thackeray would have approved ('there's no good novelist who isn't a good gossip') as Matt and Flo nurtured it with their tattered copies of the *News of the World* and anecdotes of Dorchester House: the night Lord Louis left the front door open and a candelabra burning on the step; and the story of a penis on a statue dusted by housemaids till its glow embarrassed every eye.

Trivia can be priceless trinkets, but finding me deep in the *News of the World* Mariamne thought to improve my mind. She did not realise that by introducing me to *Myths of Ancient Greece* she was only substituting the gods' adulteries, feuds, and murders for those of the wider-ranging British social system.

The Better Classes – the Lower Classes; the English – the Australians . . . Generalisation and juxtaposing are tempting games. What makes a situation funny to an Englishman and cruel to an Australian? Perhaps the early days in Australia were too brutal to encourage a cutting wit. The free settlers seem to have been industrious worthies rather than wits. Even the Irish left wit and imagination behind them in Ireland, bringing with them their peculiar brand of Catholicism, their cantankerousness, and their love of booze and racehorses. I see Australian humour as pretty woolly till the middle of the Twentieth Century. It seemed non-existent when I returned to Australia after Hitler's War. Then as cities grew, and with them a new generation, an abrasiveness began to creep in, an increasing sophistication as younger people travelled, and the outside world rubbed off on us. Even so, satire remained suspect if aimed at 'warm' Australians, the majority of whom are not the golden-hearted beings they would like to think. All of this is understandable. Black humour and satire germinate in wars, disillusion, and hunger. So it throve in Berlin, and there has always been a vein of black in the British people's most extrovert slapstick, derived from hunger and exploitation in the past, and in the present century from close involvement of the civilian population with the two great wars.

So I think the humour I most appreciate derives from those bitter days when I was a despised Colonial attempting to pass myself off as an English schoolboy, then hanging round the theatres in the 'Thirties, the London streets during the Blitz, and the long boredom and isolation of war in the Western Desert where most of my companions were frustrated British.

Though it is doubtful he would have turned out to be a golden-hearted Australian whatever the circumstances, all this has undoubtedly contributed to the black in White. It is also why Matt and Flo Davies continued to play such a part in his life, and why they loved and understood one another.

I remember an occasion when Matt and Flo were very old. We

33

were sitting in the fug of what had once been a gleaming, enamelled kitchen, now smelling of lack of attention, dry-rot, and half-wild cats breeding under the floorboards. A caged love-bird tinkered with his toys or looked at himself in his little mirror. We had been discussing the case of an unfortunate abducted woman who, it was believed, had been fed to the pigs on her captors' farm. After thinking it over behind his grave, comedian's mask, and rearranging the arthritic's aluminium crutches, Matt came out with, 'Put you off a pork sausage, wouldn't it?' The three of us laughed, black laughter for a bleak but hilarious world which we understood in our bowels, as warm, deluded Australia could not.

I was born with that silver spoon, the metaphor popular novelists and Sunday journalists love to trot out. Unfortunately, or not, I was given eyes, hyperactive emotions, and an unconscious apt to take over from me. My childhood should have been a happy one, and was when anxieties and foreboding did not bring on asthma, that curse of half my life. Adolescence was probably not much different from the unmade bed in which every young person of average sensibility tosses and turns interminably. In early manhood I began to see that the external world was no other than the dichotomy of light and darkness I sensed inside me. The principles instilled into me by my parents and Lizzie were only related to my rational self. My mentors could not have imagined any of the darker undertones, or the implications of deviant sexuality. I have stuck by most of their principles while knowing in my irrational depths what it is to commit a murder, or be murdered.

I can't remember being much worried by evidence of sexual ambivalence. I indulged my sexual inclinations at an early age. What disturbed me was the scorn of other boys, not for my sexuality, which they accepted and in some cases enjoyed, but for a feminine sensibility which they despised because they mistrusted. It is much the same situation when predominantly masculine men despise women for subtleties the male lacks, while making use of their sexuality.

In my case, I never went through the agonies of choosing

between this or that sexual way of life. I was chosen as it were, and soon accepted the fact of my homosexuality. In spite of looking convincingly male I may have been too passive to resist, or else I recognised the freedom being conferred on me to range through every variation of the human mind, to play so many roles in so many contradictory envelopes of flesh. I settled into the situation. I did not question the darkness in my dichotomy, though already I had begun the inevitably painful search for the twin who might bring a softer light to bear on my bleakly illuminated darkness. Looking back, the two languages I read at Cambridge were in some sense complementary, and part of the blundering search for a means of self-expression and fulfilment.

I had begun French early, earlier even than the orgy of mirror-watching phonetics in which a progressive headmaster met his downfall. My first teacher was a large, comfortable Swiss from Neuchatel who had settled with her husband and family in what was then Woolcott Street. (Some years later the name was changed to Kings Cross Road in an attempt to clean up its reputation.) The most fascinating part of the French lessons with Madame Henri was my dawdling approach from Roslyn Gardens up Woolcott Street observing the early risers in their lace gowns and boudie caps lounging on their porches or sweeping them after a fashion while discussing details of their trade with a colleague. What I sensed intuitively was fleshed out from the guarded remarks of virtuous adults, none of whom knew what I knew. As I walked past the terrace of sleasy houses in the early light I often experienced a jealous thrill, half fear, half possessiveness, not unlike my adult reactions to a scene my unconscious has created for me. Sometimes when I had stared too hard and too long from the opposite pavement the whores would start shouting obscenities. I must have looked a prissy child, too well scrubbed, too well combed, from a materially protected world. Their contempt would have been justified. But I mooned on as though unaware, just as the whores were unaware of the image they were helping crystallise.

For my lessons with Madame Henri we would sit together in front of a Nineteenth Century colourprint with agreeably surrealist overtones, engraved on my mind as 'The *Qu'est-ce que c'est que ça* picture'. In it stood a French Château with fish-scale roof, turrets and bull's-eyes, balustraded steps descending to a lawn where pieces of furniture were exposed for identification. 'Qu'est-ce que c'est que ça?' Marie-Thérèse intoned, pointing at an object with the tip of her sharpened pencil, and I would reply automatically, 'C'est l'armoire', 'C'est la pelouse', 'C'est un petit chien', as the case might be.

Monsieur Henri ('Charr-lee') was vaguely elsewhere in the flat. He made excellent biscuits which my mother used to buy. But it was Marie-Thérèse who kept the pot boiling. Their three boys were all in the Scouts. My favourite, who was also the handsomest, drew me a map of the North American continent, the outline in India ink, surrounded by blue-pencil oceans. The map, and a Presbyterian prayer-book in disinfectant-pink paper cover given me by Lizzie, were my most cherished possessions till Sue tore them up.

Before leaving Cheltenham I enjoyed a brief flirtation with France and the French language when I was boarded with a family at Dieppe during the summer holidays. If I describe this first real contact with France as a flirtation it is because I think a passion for France can only develop with maturity. As I see it, Germany has, or had, a mental climate to appeal to romantic, turgid, confused youth. At Dieppe I learned little French owing to the presence of my schoolfriend Ronald Waterall and a troupe of rumbustious Swedish girls all of whom spoke fluent English, though Grandmère Vincent made me copy out *les synonymes*, endless lists of them, and get by heart the poems of Lamartine, whose hand she had shaken when she was a girl. The lessons were conducted beside an enormous bed, above it a crucifix and sprig of yew, Grandmère's stomach rumbling with *tilleul* as she recited the lines of *Le Lac* in appropriately respectful tones.

In addition to this I started reading *Madame Bovary* while stretched out on an uncharitable *plage*. My inadequate French and the pebbles under my vertebrae made it torture. In the same surroundings I grappled with Colette (*Chéri*) and various novelettes, one I remember by someone called Gyp. It was all very stimulating and at the same time frustrating, not to be able to unlock a language with so many treasures behind its bars. I squirmed on those pebbles as a novice fakir must writhe on his first bed of nails. Nor could I appreciate the almost permanently grey stretch of water in front of me. None of its impressionist translucence, its veils of Proustian significance, the historic emotions with which that narrow sleeve has been washed, suggested themselves to my ignorant schoolboy eyes.

What perturbed me in the cold wind blowing off the Manche was that the oil with which I had anointed my goose-pimpled skin, and which spattered and fingerprinted my Flaubert, was not living up to the promises on the label. I began to realise I was being dyed, not tanned, and humiliation was unavoidable when the dye ran down my collar on a hot night at the casino opera where I was undergoing a crash course in Massenet – *Werther*, *Manon*, *Thaïs*. (Ah, that *Méditation*!)

I was at the right age for seduction by the facile and the frivolous. We had as inspiration the bright adolescent adults who careered about London in the 'Twenties, and though we had never been closer to them than on the pages of a magazine, they waved, smirked, and gnashed their way into our youthful imagination. So our *petite bande* roamed screeching through Sickert's mud-coloured streets, ignoring the disapproval of shopkeepers standing in doorways, and the even more shocked sensibilities of those thin French cats flattening themselves against the merchandise. We sat chattering at our favourite café, smoking *Gauloises* with queasy nonchalance, while floating apart on waves of *Pernod fils*.

All that I was undergoing, and the desire to boast when I returned to school, persuaded me I should fall in love with some pretty girl. I picked on the most voluptuous of the Swedish *pensionnaires*. She was blonde, dimpled, plump; her eyes almost disappeared when she smiled. I really fell, but in a worshipful kind of way. We embraced in doorways and on the sea-wall. I bought her a necklace of pink beads and a pink silk scarf. She must have found me amateurish, heavygoing. She could have been sleeping with an older man who was my friend, an army officer with tales of China, also boarding at 'Les Aubépines'. I was told she was carrying on for certain with a French youth typical of those who haunted the tennis courts of watering places during the summer season, all bounce and casual foulard, sophistication letting them get away with acne and hollow chests. Although my love affair was a planned one, the pangs I now suffered were those of a genuinely jealous lover. Or was it wounded pride? After parting at the quay, and the brief exchange of pathetic letters, I never heard what became of her. I expect she grew stout and Swedish, perhaps a little slatternly. I remember that grubby shadow on the Achilles tendon.

As for my friend the army officer, who had served in China and fucked Chinese girls sideways, he stayed on at '*Les Aubépines*' after I had left. I didn't hear what became of him either. In this case there was no exchange of letters, though if I had been honest I might have admitted that I was more attracted to him than I was to my flirt, the plump Swede he had probably enjoyed. His rare attempts at French whenever Madame addressed him at table, were always preceded by 'Mais oui . . . Mais oui . . . ' Perhaps I should have assaulted him in French. He had the lean hungry look and a chest protected by what appeared to be a burst hair mattress.

Germany came later. I only decided for the German language the day I arrived at my Cambridge college. I had passed the entrance exam on History, then felt as the time approached I could not face writing yet another History essay and condemnation of my fanciful style. I was confident I could rustle up enough French for the tripos. But German? From my smattering of rudimentary school German I had to advance to honours standards. I was not all that worried. After two years jackerooing in Australia I was experiencing the transports of sinking my roots in richer soil. I knew I hadn't a scholar's mind. Such as I had was more like the calico bag hanging from the sewing-room door-knob, stuffed with snippets of material of contrasting textures and clashing colours, which might at some future date be put to some practical, aesthetic, or even poetic use. I believe it is this rag-bag of a disorderly mind which has more than anything offended some of my Australian academic critics. For them the controlled monochrome of reason, for me the omnium gatherum of instinctual colour which illuminates the more often than not irrational behaviour of sensual man.

Between '32 and '35 I spent the greater part of every vacation in Germany. I had to catch up on a language which, until I visited the country, I could never take seriously: comic, hedge-hog words constantly colliding, syntactical structures to get lost in. Then while exploring the country itself I became obsessed by its Romantic literature, which in later life was dispossessed by my passion for the French. I moved drunkenly through German landscape, the feverish greener-than-green which so often gives out a smell of rotting, the black-green of pine forests, the austere, dead-green North, scruffy with sand as it approaches the Baltic coast. In Hanover, the city with which I became most familiar (later the *Heimat* of Voss and Himmelfarb) I was cosseted by a homely *Bürger* family, who put me to sleep under feather beds, and fed me *Kartoffelpuffer*,* *Häringsalat*, and *Wurst*. I loved it all, and them. I returned over and over to the Oertels, until human relationships crumbled under political pressure.

But in the early stages of my love affair with Germany, in spite of some disturbing personal clashes with the Nazi men-tality, the burning of the books while I was in Heidelberg, and shadowy, persecuted Jews living along the same street, Hitler was still an object for scorn and cynicism among liberal-think-ing Germans. We laughed our heads off drinking *Schnapps* on the safe balcony above the Holzgraben in Hanover, and Lotte overpainted her lips in defiance of the tin sabre.

During these years of political ferment, of which I grew aware, like many uncommitted Germans, only when my nose was rubbed in it, I wandered over most of the country, not so much Germany of the period we were living in as my own romantic reconstruction of the past. Even as fantasy it was not without its sinister side. As a foreigner followed by a troupe of jeering children, cropheaded boys, and pigtailed girls in tunics of Prussian blue, I stalked along the quay at Koenigsberg pretending nothing was happening. To their mackerel eyes I must have appeared a grotesque figure. A backdrop of ware-houses, formal as painted scrim, seemed to be awaiting some diabolical performance such as the Habima might have staged as a revenge in this most Aryan of towns.

I remember visiting the little Baltic resort of Cranz, sinking up to the ankles in heavy white sand on the edges of the town, even in the streets of whitewashed weatherboard houses where the light lay as solid and golden as the amber found along the coast. Blander than Koenigsberg, Cranz was not less sinister: it belonged nowhere in time and had no connection with any country I had visited. There could have been a Russian influence. Russian families had patronised Cranz before the Revolution. Unexpectedly the streets of Abukir reminded me of Cranz when I was stationed in Egypt during Hitler's War. Abukir had that same timelessness of dream life and the smell of primitive washing spread to dry on bushes.

At Koenigsberg I put up at an old-fashioned hotel. Kant might have eaten his meals in the practically deserted, probably unaltered dining-room. Throughout the building a thick silence and smell of dumplings. On a landing a wall-mirror framed in mid-Nineteenth Century gilt showed my diffidence, still awash in the glassy shallows as it had been reflected on a smaller scale by the looking-glass in Hayley's folly at Felpham, Sussex.

Apart from Hanover, where my cosy Oertel family made all the difference, Weimar was the only town during the years of my German wanderings which encouraged a sense of well-being. A marriage of elegance and sobriety in its streets, houses, gardens and mini-palaces, proposed a celebration of rational man. Suddenly I felt at home in surroundings to which I don't belong spiritually, and in spite of my lack of sympathy with Weimar's two great poets, one a manufactory of German platitudes, the other a genius who founders in his hypocrisy and pretensions as a human being. For me Tolstoy is the only literary genius who survives his own hypocrisy. Not Goethe. I am almost persuaded when I read Eckermann's description of his idol lying dead. Again and again in the course of the *Conversations* he almost brings it off, but at the end one realises it is the humble acolyte, himself an original and erudite mind, who has infused the padded monster with his own compassion. (Thank God for the minor characters, like Goethe's Eckermann and Willie Yeats's miraculous father, J. B.)

Never very warm towards me, France grew colder during my trafficking with Germany. When I returned to France I was told I spoke French with a German accent; just as at school in England I was accused of being a cockney or colonial, and back in Australia, 'a bloody Pom'. Language troubles have widened the split in my nature. This was not reduced when in Danzig, while it was still a 'free city', I asked the way of a passer-by and was told, 'Sorry, I too, am a German.' Admittedly my vanity preened itself a little on this occasion after I had recovered from the shock.

Of course I am vain, less since losing my teeth and to some extent my sight. If I have not yet lost my mind, I can sometimes feel it preparing to defect. But vanity is deeply rooted in me. I am said to have come home from my first morning at kindergarten and announced to Lizzie, 'I've got the best voice in the school.' 'How do you know?' 'I've heard the others, and I've heard my own, and I know which sounds best.' I often wish such conviction had followed me through life.

I can't remember vanity in Dick; he was a plain good unimaginative Australian male, the kind that panders to pretentious wives. Ruth was vain, but her vanity was easily deflated; like her son, she suffered from diffidence. Though she despised the Whites for being rich, she enjoyed the advantages money brought. She would tell us how hard she had worked at 'Piercefield' during a far from affluent youth. We had never seen her work, except to get into an apron at Mount Wilson before she lost interest in the place, and stir a cauldron of plum jam. (After the jam had calmed down we were given the two-shilling pieces said to keep it from burning.) We had never seen our father work. Marrying at forty-two he retired from the land and started leading an urban life when Ruth refused to live at 'Belltrees' ('too many brothers- and sisters-in-law'). I think I might have had greater respect for both my parents if I had seen them working. All I have to go by is a snap of Dick, an honest, boyish, burnt face running with sweat as he squats during a spell in the shade of a thinly furnished gumtree, his horse's nether lip drooping a couple of inches above his shoulder. The bland blue eyes never let the camera down, nor anybody else. I remember him returning from working visits to 'Belltrees', and their other station 'Tereel' near Gloucester. The smell of his valise and oilskin invading the hall challenged the furniture to convince. Our mother had put on an apron to welcome our father to a meal she had not cooked. More than anything I was fascinated by the leggings covering his stocky calves.

I admired rather than loved my mother. I loved to follow the rites of the dressing table, the robing and disrobing, or as much of it as I was allowed to see. Lipstick was still 'lip-salve' and hidden under the handkerchiefs. She was not a woman given to scenting herself. There was the more innocent smell of powder, the dramatic stench of singed hair when the crimping-iron had been overheated, the mysterious perfumes of dresses hanging limp in the wardrobe waiting to be brought to life. I think she enjoyed my presence, like an actress with a safe audience of one. I stimulated her vanity and some of it surely rubbed off on me.

Vanity, that paper edifice quickly brought down, and as quickly re-erected . . . I can remember my mother appearing in a mountain court as witness to a motor accident, yellow and wrinkled, voice thin, de-fused, an old woman in a morning. I have remembered this occasion during some of my own forced appearances in public – reduced by dread, shorn of all imperiousness, till I started what Lizzie had always warned against, blowing my own trumpet.

Growing up in a period when drabness was expected of the male sex, my vanity could not express itself through dress. When it finally became possible I was already too old; there's nothing so pathetic as trendy old men and the lipstuck mummies of women. Instead I suppose I've indulged my vanity by tricking myself out in words. Not all ornamentation. Part of me is austere enough to have conveyed the truth, I like to think, but that again could be vanity. If I believe this today, tomorrow I may feel that truth is the property of silence – at any rate the silences filling the space between words, and over those I sometimes have control.

'He didn't get it from the Whites!' One of the 'Saumarez' spinsters was very firm on that score. She confessed they had bought my first novel; it was somewhere in the attic. One of the 'Belltrees' cousins looked me up on a visit to London when I was living in Ebury Street and had started writing. We spent an agreeable evening. At the end of it she said, 'It's been interesting to meet you. We thought you were kept over here because you were mad or something.' My parents were so disgusted they never spoke to her again. Yet a few years back, driving in a taxi past Selfridges, Ruth had flown at me because I had refused to go with Dick to a cricket match at Lords'. 'I never thought I'd have a *freak* for a son!' As Suzanne may have been the prototype Australian Sue, Paddy was probably its first labelled freak.

The White dynasty at 'Belltrees' began with the migration of Henry, Ernest, Arthur, and Victor from 'Edinglassie', leaving that property to James. Ernest of the sympathetic face I never met. A middle-aged bachelor, he died of peritonitis on the train to Sydney not long after my parents married. An elderly spinster acquaintance once told me he should have been hers, but became involved with a Muswellbrook 'seamstress'. Ernest, Arthur, and Dick all performed on the polofield. Henry didn't. The 'Belltrees' brothers had to rope in an outsider to make up their team.

Henry was the most forceful, literate, authoritarian of the brothers. The siblings might have seen him as freakish if they had dared. Ruth disliked Henry because, she said, he treated his wife abominably. He was certainly eccentric by White standards. He collected early Australian stamps, sheets and sheets of them, which he gave to the Mitchell Library, Sydney. He employed a team of men to scour the country for ornithological specimens; the birdskins and a collection of eggs he gave to Melbourne (Sydney was rude to him). He had a library of Australiana. There were also books the Whites would have found embarrassing. One of these turned up in South Australia in the library of the Dutton family. Geoffrey Dutton showed me it. The tone was faintly pornographic. One illustration showed a little pug jumping over the ankle of a naked Edwardian beauty.

As I remember him my Uncle Henry was a short man of fiery complexion, his eyes as cold as blue glass until they blazed with enthusiasm or anger. In spite of my mother's dislike for him, he was the only White uncle I liked, and he seemed to reciprocate, perhaps sensing a fellow eccentric. He used to give me presents of stamps to encourage the philatelist in me. I tried to live up to his hopes, but my interest in stamps withered after a little.

Both Uncle Henry and that dim hour-glass, the abominably treated Aunt Maud, gave me pocket money to take back to school. I can't remember more about Aunt Maud than the pocket money and her silhouette. Milly her sister I got to know better, but only after Arthur died, when without his restraining influence she found she liked me.

The brothers appear to have competed for Dick's affection. In his last years, Henry used to write to him every day, sometimes twice, copying each letter into letter-books which can be seen today in the 'Belltrees' library. I don't remember my father talking much about Henry. Perhaps he didn't dare in front of Ruth, whom Henry wooed unsuccessfully with presents of hare and eel. She was one of the first Sydney gourmets. She hung the hares in the garage till the cook threatened to give notice. ('But it isn't ready till it's soft enough to eat with a spoon!') In spite of these delicacies my mother never relented in her opinion of her brother-in-law.

Dick was said to be Arthur's favourite brother. They went about together when Arthur (and Milly) came to Sydney. Dick always paid, our mother told us. When one of Arthur's horses won a fat purse in an important race he offered Milly a present of her choosing. After knotting her hands and protesting she asked for a flushing lavatory. Their home, a short distance from the 'Belltrees' homestead, was a discreet red-brick cottage, 'Kyoto'. If the cottage looked suburban, the homestead with its wrought-iron balconies and formal flower beds, sunshades protecting roses and chrysanthemums, was more like the city mansion of some Edwardian financier. It went with Henry, I suppose, as the cottage suited Arthur and Milly, who were in everything discreet.

The Arthurs enjoyed practical jokes of a mild kind, lavatory seats and clothes-brushes which played a tune. They loved brass bands. If they went to the theatre, jugglers and trick cyclists were what appealed most. Arthur was a slow, wearisome punster. But he had a profile, and a slim figure which impressed on horseback, particularly a sidling horse. He was given to wearing white ties against a white shirt. As far back as I remember his hair was thin and silvered. He had a thin silvery moustache well adapted to his thin lips, which opened only just enough to let a thin voice give utterance. His voice was pale yellow, I used to think, as he and other men of that day droned on through the dust and heavy Australian light about country matters which concerned them.

Arthur enjoyed a cigar, while ignoring the possibility that others might have accepted one. He and Milly followed the cricket from hemisphere to hemisphere. They sat in the dark to save the electricity. In his old age, when they were living in a suite at the Hotel Australia, he was taken to hospital suffering from malnutrition; their staple was baked custard.

Milly was small and thrush-coloured, an apologetic smile in an incredulous face. I think life and Arthur were too much for her. In different circumstances she might have turned out differently. She told a friend they didn't like Paddy because he was clever. Yet she took to visiting us regularly at Castle Hill after Arthur's death, where she seemed happiest tittuping about the garden picking nosegays of pink oxalis to take back to her hotel suite. In Arthur's day she had done what laundry she could in their bathroom. There were always handkerchiefs stuck drying on the mirrors. And flowers dying in vases. And old voices trying to disentangle themselves from the illogic which had become their vocation. When Arthur was carried off to hospital for the last time, he gave Milly strict instructions on how to get there by tram. In addition to his share in 'Belltrees' which went to those who continued farming it, he left half a million pounds to institutions like the Kings School Parramatta, the Salvation Army, and the Church of England.

Milly lived on with a couple of nurses in attendance in the same suite at the Australia. After some persuasion, and doubts as to whether she ought, she bought herself a radio, and would

ring me up late at night to tell me she had been listening in to
Parliament.

One thinks to escape, but doesn't, or not wholly: the finger-
prints were taken early on. The past recurs in the dressing-table
mirror, wisps of it in benign dreams, or those darker ones in
which unfulfilled, half-forgotten lusts explode. Worst of all are
the consciously created fictions, because concrete evidence of
what one has not dared admit.

I grew conscious of wanting to be a writer on leaving my hated
English school and returning to the Australia I had longed for.
No, it wasn't so much a case of growing consciousness as a
matter of necessity. Surrounded by a vacuum, I needed a world in
which to live with the degree of intensity my temperament
demanded. So, while jackerooing, first a year in the Monaro,
then a second at Walgett with the Withycombes, I wrote three
rambling immature novels, fortunately never published,
though I was depressed and mortified by their rejection. Bits of
the first two surfaced in later work; the third gave me the
foundations of *The Aunt's Story*.

I was alarmed at first, then permanently unhappy, to return
home and find myself a stranger in my own country, even in my
own family. I had been released from prison of course, but
freedom was not what I expected. (I was too young to know that
freedom is a theory anyway.) My gaol sentence had left its
stamp; I met with suspicion and often undisguised dislike. It
was a situation I have experienced visiting countries like
Czecho-Slovakia and Turkey without having a word of the
language. In Australia the difference lay in my understanding
every word the natives spoke, while they cottoned on more or
less to the diffident mutterings of the foreigner, without our
communicating to any but the most rudimentary degree.
Throughout childhood and youth I had received warnings of the
price I would pay for being different: the overheard remark that
my parents had got a 'changeling', the jeers of English schoolboys for a Colonial in their midst, my mother's cry when she

accused me of being a 'freak'. But none of it prepared me for the shock of returning to Australia after school. In the four years I spent away my imagination had created an ideal land from which my mother's ambitions had cast me out. However kind, generous, affectionate my family were on taking me back, they still did not understand the peculiar youth who had developed out of their difficult child. Schoolfriends had dispersed and were working in different parts of the state. The few I met up with were daunted by my English inhibitions, which must have appeared as coldness, and of course there was the eternal barrier of speech. A film gathered on the eyes of those faced with my accent, whether pretentious hostesses who fancied themselves as sophisticates, or humble innocents like Harry England who trained my father's racehorses. I had known Harry since I was a small child when we used to go out to look at the horses after Sunday dinner. Harry would go through the routine of running a trembling hand over swollen fetlocks, explaining why a horse wasn't up to form. He had a spongy, respectful upper lip, very obviously shaven for our visit. If I was fond of Harry, I found his wife fascinating. She would entertain us to strong tea and floury scones inside half-drawn holland blinds and a drone of blowies. She was usually wearing a pair of her husband's long woollen underpants above carpet slippers. Clusters of little blonded curls jiggled on either side of her face, and she smiled and smiled, a china smile. In their embarrassing respect for wealth the Englands belonged to a race of Australian sycophants now extinct. Not that we don't have sycophants today, but they are cheap and tinny by comparison; the upper lip doesn't tremble in the same affectionate way. Faced with what I had become at school in England, Harry could only tell me that I was 'the real gentleman'. He had no idea how he cut, and set me back.

I badly wanted to love someone. There were the girls my mother produced, and whom I met with relentless regularity at the dances I was forced to go to. Brittle, often pretty creatures, they twittered about horses they had backed that day at the races, friends who had just become engaged, and the most recent Garbo film. One girl the others pointed out with grudging awe was said to have read 'the whole of Trollope'. Though scenting another outsider, I shied away; my chemistry lacked some essential property. We got to know each other only in our middle age, and enjoyed a good, though superficial relationship.

Which of them do you like? was Ruth's recurring refrain. She desperately wanted me, not so much to fall in love, as to become engaged, so that she could stage-manage a wedding. She wanted me married, like every grazier, banker, lawyer she had ever known. She wanted Sunday lunch at the Royal Sydney Golf Club with a daughter-in-law and grandchildren.

In spite of her taste for exotic foods (for which I have always been grateful), the latest, and sometimes daring clothes, and the determination to go one better than any other Australian family by sending a son, not only to an English university, but an English public school as well, Ruth was at heart a conformist. To be admired for originality was not the same, finally, as to be congratulated for one's contribution to family virtue and continuity. She was bitterly disappointed, poor girl. I think unconsciously she blamed me for turning her into an outsider.

When I say I didn't love my mother, or only admired certain of her traits, Manoly points out that I wrote to her every week. Why not? I had a sense of duty towards her, and I've always been a creature of habit. It might have helped had I been able to love her on returning to Australia after school, but at that period I don't think I even admired her; I was too harassed by the hysterical refrain 'Which one do you like?' to see her as more than a predator descending again and again to tear out an essential part of me. Lizzie was too far away, minding her own business on the mountain. Sue and I still sulked and fought. I can remember being scornful on catching sight of a flea from one of the dogs hopping out through an embroidered eyelet in her dress as we sat side by side on the sofa. My desire for the ideal, the statues embodying perfection, was such that a miserable flea could threaten a human relationship.

Had I been able to talk to him, and if, at the risk of sounding priggish, there had been some vaguely intellectual ground on which we could have met, I would have loved my father. I expect it would have led to more complicated overtones in my relations with Ruth. I could not admire Dick enough to love him; innocent goodness, generosity, kindness, were not enough. I loved servants like Lizzie, Flo and Matt through their connection with everyday reality. I had never seen my father in the context of reality. He had withdrawn from it before I was born. Now if he tried to talk to me it was about cricket or football matches or the performances of racehorses. I aimed at escaping from any room in which we found ourselves together. Since then I have sometimes wondered how much suffering I caused him by surrounding myself with what he must have seen as mysterious affectations, a world of unreality which the literate find in books.

Even his size was against him, and the blue, shadowless eyes, the quiff of hair above a bland forehead. His skin was burnt and wrinkled as far back as I can remember. I was repelled by and attracted to his hands. If I could have brought myself to touch the back of one of them the dam might have been broken. It could either have solved our problem or caused worse disaster.

As I could not come to terms with the inhabitants, either then, or again on returning to Australia after World War II, I found consolation in the landscape. The ideal Australia I visualised during any exile and which drew me back, was always, I realise, a landscape without figures. The actual, noble, though often harsh and bitter Monaro scene was my spiritual sustenance in the year I spent working there. I would be trying to convey something of this in the juvenile novels I wrote by lamplight alone in my room, till the blast of community singing from the wireless in the kitchen drove me out. The scent of wood-smoke, drool of hens roosting in the two-seater dunny, cat or dog hardly more than shadow brushing up against my legs, helped exorcise the community voice, as chattering with cold, I climbed the hill towards the iron water-tank, below us the silver river winding through a plain of white tussock.

At Walgett my uncle and aunt ignored the community voice. Instead there was Clem's telephone voice, conversations with a mate which lasted hours at a time on a party line, after the meal had been cleared away, while I wrote my obsessive novels on the dining table. Again landscape played the major part in my life, the splendid sweep, the cairns and monoliths of the Monaro replaced by dead-level plains with their skeleton trees, clouds of galah and budgerigar, a formation of creaking pelicans, soil white and cracked with drought till transformed by rain into black porridge, giving way to an inland sea, on which was laid, in a final act of conjuring, an emerald carpet patterned with flowers. At Walgett I experienced every possible seasonal change and corresponding changes in myself.

My year with Clem and Margaret was far more congenial than the one before at Adaminaby. I had visited them as a child. My parents used to put me on the train at Sydney Central, nominally in the guard's charge, but with all kinds of additional instructions should he fail me. I would spent the day at Maitland with the great-aunts, in their little house like many others, brown, blistered weatherboard, and a back yard in which the sun-baked silt from floods was fitfully shaded by a pepper-tree. In the afternoon Aunt Lucy who hated the Micks took me by steam tram to Newcastle, where we visited a relative, a dignitary of the Anglican Church. At night the aunts put me on the train for the north, again with warnings should the guard neglect his responsibilities. The most frightening part of the expedition was changing trains at Narrabri by starlight. After that it was plain sailing through the dawn, and an endlessness of dust and mirage, with wild melons embroidering the edges of the railway track. A dinner of tough mutton and floury gravy at Burren Junction. Then Walgett, always later than schedule. My dark, dictatorial uncle met me with the buggy. I felt as proud as he looked, and on reaching the outskirts of the town he allowed me to take the reins and drive. It was my moment of stardom.

From these childhood visits to my uncle and aunt I knew their household and their habits when I went there to work as a young man. Childless themselves, they were somewhat diffident of the young, afraid I might poison myself with slot-machine chocolate while in their charge at Coogee Beach, or far worse, fall for a girl my mother might disapprove of when I reached the age of indiscretion. Margaret was one of the slave wives of the period. I was always very fond of her, but did not appreciate her till in my middle age I understood her remark about the treadmill. She was always preparing food and clearing away the crumbs. Without telephone interruptions or fire in the kitchen, Clem and Margaret and their overseer took a quarter of an hour over a meal. When the heat was at its most intense under the iron roof Margaret would dip her nightie in water (mostly in short supply) and stretch out on the lino. She had a storeroom full of waterglass eggs and meat in pickle, and a little parlour where she read her library books from Sydney. At that time her favourite author was Ruby M. Ayres.

As a character, Clem was far more spectacular. I admired him as an authority on country matters and for his physical panache. I didn't love him, because it wasn't done to love a man except after a wooden, apologetic fashion, but in any case, as a man, he repeatedly let me down through exhibitions of violent Withycombe temper. I remember one occasion in particular when I failed to steady a nervous horse whose fetlock he was treating. Our relationship was temporarily destroyed when the man I admired became almost jellified with rage.

So again I preferred a landscape. It answered my needs. More passive than the Monaro, it was also more sensual, sympathetic towards human flesh. Perhaps because a rare commodity, water played a leading part in my developing sexuality. I was always throwing off my clothes to bathe, either at the artesian bore during a pause from mustering, the water ejaculating warm and sulphurous out of the earth, or in the river flowing between the trunks of great flesh-coloured gums, to a screeching, flick-knife commentary by yellow-crested cockatoos, or at night in the hollow below the homestead if a good season had turned it into a lagoon. Here I was joined by the men who worked about the place, whose company I enjoyed without quite becoming their equal.

The way to the lagoon was stony. I once found a pair of old high-heeled shoes amongst the junk dumped in the bathroom. I wore them, tottering across the stones till reaching the acquiescent mud, the tepid water of the lagoon. My companions turned the shoes into a ribald joke, acceptable because it was something we could share. We continued joking, to hold more serious thoughts at bay, while we plunged, turning on our backs after surfacing, spouting water, exposing our sex, lolling or erect, diving again to swim beneath the archways made by open legs, ribs and flanks slithering against other forms in the fishy school, as a flamingo moon rose above the ashen crowns of surrounding trees.

Abortive novels are as painful as failed love affairs, but have to be gone through. Living in London, trying to justify an otherwise useless existence by being a 'professional' writer, my failures were the more abysmal, my few successes, though exhilarating at the time, no more than false starts. Those early effusions by lamplight at Adaminaby and Walgett were more honest in their lumbering after truth than my subsequent chase after a fashionable style in London, my frustrated longings for sexual fulfilment more deserving of reward than the conscious traps I began setting for love, and which usually went off at half cock.

After coming down from Cambridge I took a bedsitter in a lodging house in Ebury Street to which I was introduced by my schoolfriend Ronald Waterall. Ebury, and its cross-street Eccleston, with nearby Victoria Station, were the microcosm to which I returned at any phase of my London life. Those with pretensions liked to think of it as Belgravia, but the realistic knew it to be Pimlico. Ebury Street with its bedsitters, flatlets, small businesses, and round the corner a post-office, could not afford airs. Architecturally it was unassuming. On Sunday mornings the yellow-grey-to-black façades of the houses suggested a cheap junk-shop print; as we who lived in it could only be seen as a junk lot: aspiring or failed actors and writers, decayed gentry (the Belgravia touch), and raffish often drunken members of the nobility inclined to dance in the street without their clothes. Yet Mozart had lodged in a house at one end and composed his first symphony there during a concert tour as a boy. (For a time the same house belonged to that genuine eccentric Lady Sackville. It became a brothel, and was hit by a bomb in World War II.) George Moore had lived in Ebury Street (there was a Wedgwood-blue plaque to confirm it) and on the tinsel side, Noel Coward and Godfrey Winn.

I loved every brick of it. At last, I felt, I was into life, in a green pork-pie hat and a black polo sweater.

Ronald Waterall I came across my first day at Cheltenham in the house of our imprisonment. He made his entrance camping down the sweat-room stairs singing,

'California here I come,
Way back where I started frum . . . '

Such flamboyant disregard for convention cheered while also frightening me a little. He was an angular boy with spectacles and flaming hair. After we got to know each other he never stopped taunting me for being mouse-coloured and devoid of personality. In spite of feeling depressed by his opinion, a friendship had to develop because we found we were both stage-struck. We started writing to actresses asking them for autographed photographs, giving our address care of a book-shop in the street along which we walked from house to school. Our trafficking was considered scandalous by any boy who found out about it. We were asked, 'Do your people know?' In fact they did, which made the situation more outrageous. Some of the others referred to us scornfully as 'the Chorus'.

Because Waterall was a couple of years older than I, we had to ask permission to walk together. The boys of Cheltenham College were expected to link arms and walk by twos and threes, I imagine to show how well we all got on with one another. Surprisingly, I was allowed to link with my older friend. I never succeeded in working out how it came about. I was even allowed to visit him in his study, for an hour on Sunday afternoons, after he had graduated from the sweat-room. Though the house-master would burst in to see what we were up to, our immorality only extended to sorting our autographed actresses (quickly hidden) and to playing numbers like *Valencia* and *Tea for Two* on the portable.

In the hols we really went to town while staying with Granny Waterall in Warwick Gardens, Kensington. An old lady in an auburn wig, Granny Warts suffered from bronchitis which kept her shut up most of the year and left us free to entertain ourselves during the visits. This was fortunate because Granny disapproved of theatricals, unless a musical like *Lilac Time*, or something with a religious theme. The film *Ben Hur* became our alibi while we spent afternoons at plays running only on their promise to shock. I remember one, a Mayfair version of Potiphar's Wife, in which the leading lady, we were told by the picture papers, wore transparent pyjamas while trying to seduce her chauffeur. Whether the truth lived up to the promise, there was always the dusty glamour of the stuffy theatres, the clatter of tea-trays, the superior programme-sellers with their black bows and modulated voices, the rows of spongy anonymous faces, and the knowledge that we shouldn't be there. As the lights faded, the curtain stirred to release the first scents of painted and powdered flesh, the first glimpse of a room which, in its artificiality, was more real than any room we knew, the actors with their exaggerated stagey accents more desirable than actual people.

We must have spent hours hanging round stage doors waiting for the stars to come out. If when they appeared they looked drabber, sweatier, larger-pored, than we knew them to be in actual fact, it did not destroy the illusion. On one occasion we waited so long outside the Coliseum for the Pounds Sisters, Toots and Lorna, Ronald had to go in search of a phone to assure Granny that *Ben Hur* had lasted longer than our calculations. The worst happened of course. Toots came out while he was away. For months he couldn't forgive me my brief and breathless intercourse with Toots while he was perjuring himself in the telephone box.

Life grew more sombre after Waterall left school and was swallowed up by British Tobacco. By then I had a study of my own, and had graduated from *Valencia* and *Tea for Two*. Instead my portable ground out *Chanson Hindoue* and *The Death of Ase*, and I had developed my passion for Chekov, Ibsen, and Strindberg so much deplored by the housemaster. I corresponded with Ronald, but the link was cut after he wrote describing the visit of one of our contemporaries to a brothel, and worse still, informing me that he had heard of a certain prefect's crush on a younger boy. Rumours flying round the house reached the

1 Fashionplate photograph for *Big Toys* programme.

2   The White brothers. Front row, left to right: James ('Edinglassie',
Muswellbrook), Frank ('Saumarez', Armidale), Henry ('Belltrees',
Scone). Back row, left to right: Victor (Dick), Arthur, Ernest (all of
'Belltrees').

housemaster. In stamping out corruption he put an end to our correspondence.

Without a lifeline I took to solitary walks. I even became a long-distance runner, and didn't do so badly at it, well enough to be congratulated by one of the prefects I most loathed. I believe he became a general. I wonder whether that narrow, almost fleshless, borzoi skull is still above ground, or whether it lies whitening, snapped shut on the last of its vicious intentions.

By the time I settled in Ebury Street Ronald Waterall was a refugee from British Tobacco. He had become an actor, Ronnie Waters, in spite of his parents' disapproval; collecting actresses' photographs was good for a laugh, to confuse theatre with the business of life was another matter. Ronald was a bad actor. Unlike most others, he finally realised, but at the time when I crept mouse-wise on the London scene he was still in the grip of his illusion, playing juvenile lead where anybody was wanted for tennis, from the Isle of Wight to Westminster, or hearty young, pipe-smoking fathers in plays for children, or singing the odd number in topical revue. The mouse went along with it admiringly. What the basically conventional actor did not understand, any more than the mouse itself, was that other, psychedelic worlds were preparing to revolve in the mouse-head.

For the time being I enjoyed sharing the preoccupations of a group of small-part actors and actresses who sat around in the lounge of the Arts Theatre Club waiting for the roles which eluded them, worrying about their health, their rent, their liaisons, their abortions. In love with it all I offered myself for a backstage job at the Westminster Theatre, then preparing a production of Auden's *Dance of Death*. I was asked by the director Rupert Doone whether I could do things with my hands. Unfortunately, or fortunately, I was turned down when I confessed I couldn't.

Most of the aspiring met for lunch at a little restaurant in Newport Court where two middle-aged, sweet-faced New Zealand sisters served an innocent meal for two shillings, and here the eternal themes, the elusive roles, health, rent, love, and abortion were further discussed. I was often deeply depressed as I ate my sliced hardboiled egg and lettuce ribbons which smelled of knife and vinegar, believing that of all my acquaintances I was the only one who would never enter into life or art. I grew ashamed, but I don't think anyone else was aware. Like all the obsessed, they were grateful for an audience who showed an interest in their obsessions.

Certainly there was a danger that, from constantly playing the confidant, I might never become a participator, and one superficial acquaintance expressed her concern, probably because she would have liked to initiate me into participation. Insufficiently creative herself she did not realise how intensely I was living the lives of those around me and that my creative instincts would finally push me over the edge.

The house where I lodged during this first tentative phase of my London life belonged to Inez Imhof, a Swiss from the Ticino, whose brief marriage with a German ended in his death from tuberculosis during the First World War, and her brother Mario Maranta, a capricious little baldy man who was also a great confectioner and pastrycook.

Like Matt and Flo Davies, Inez was one of the artists of domesticity. Perhaps her twangling nervous system was responsible for the subtlety of the outwardly simple food she dished up. (A certain lodger was content to eat one of her plain omelettes every evening of his life.) That Inez did twangle, everybody knew and accepted; even her hysteria was worthwhile.

Over the years I returned to the house, between moving from one flat to another, again on the outbreak of war and at the end. Physically Inez was not unlike Lizzie my nurse-mother, with Italian temper replacing Scottish discretion. When I arrived in London for demobilisation and could find nowhere else to go, she took me into her bomb-damaged house. Her basement kitchen was a warm womb in times of need, smelling of fresh laundry, and the promise of a delicious meal even in starvation days. It was always full of cats, which in the palmy era before the war would spill out and lie in skeins or heaps on the tessellated approach from the pavement to the front door. In the latter days of my lodging there, the brightness had gone. Peace did not return with the peace. The nervous system of all London was twangling, ready to go off into an Imhof-Maranta screaming match. Like many of my friends, Inez died of cancer. But in the aftermath of war, her house, and in particular the bomb-scarred version of a pre-War bedsitter, offered me a structure I built around. I wrote a play, *The Ham Funeral*, and conceived a number of fitful, ultimately fruitful ideas during a long bout of bronchial pneumonia.

After coming down from Cambridge and spending a short time in London I realised I could not go back to Australia and drag out the rest of my life on the land as originally intended. I was determined to work, at what I still wasn't altogether sure – writing? the theatre? somehow earn a living in that great metropolis where success offered glamour, and even failure was less painful for mouse-coloured nonentities. Anonymity was anybody's right in a city like London, whereas in the village life of Sydney before World War II it would not have been possible, and failure would have stunk in the nostrils of the society to which we belonged.

I wrote and broke the news to my parents. It was not a particularly brave act: I think I believed they wouldn't abandon the son who was letting them down. Always more realistic than they, Clem must have shown them that I was not cut out for the land. Fair enough, but if they accepted that, what about a doctor? or a diplomat? acceptable professions which had been discussed. An artist in the family tree was almost like a sodomite; if you had one you kept him dark.

Like the more pretentious Australian women of her day Ruth saw the arts as charming hobbies which those of her own social level might safely cultivate. A professional was something she would have found too difficult to explain away. Even after I had published several books, she couldn't really enjoy her pride because I wasn't the kind of writer she could produce at Australian cocktail parties. Her friends considered me mad, scandalous, or a joke. Setting aside family loyalty and the fact that she had given birth to me, I expect she agreed with them on all counts. She remarked after reading *The Aunt's Story*, 'Such a pity you didn't write about a *cheery* aunt . . . ' for although she remained convinced that she should only buy what was English-made, she also subscribed to the Australian myth that you must face the camera with a grin.

Poor Dick, who had never read a book unless a studbook, or one of the thrillers Ruth brought him from the library, to fall asleep over in his leather armchair after dinner, didn't know what it was all about. When, as an argument for my staying in London, I started writing the plays agents and managements kept turning down, he was impressed by the information that I had typed and bound the plays myself. I can see him holding a script in his hands on one of their visits. As an outcome he decided to give me what was in those days a sumptuous allowance of £400, so that I could stay in London and carry on with whatever I was up to. I'm sure he had no expectations, which made his gesture a nobler one, particularly coming from a White.

I moved into a minute flat directly opposite the Imhof-Maranta establishment. Waters and I could hang out of our respective windows and shout at each other, provided the gist of it wasn't too scandalous. If it was, each waited for the other to lift the phone and pay for the call. I had usually got my breakfast over, but such early morning confidences were accompanied by a munching of lettuce, crunching of toast, slurping of coffee from the other side of the street.

Originally my flat had been a room on the third floor of a smaller Ebury house. One who is nowadays called a 'developer' had partitioned it into living-room and bedroom, with a second partition converting a corner near the front door into a cupboard of a kitchen. Beyond the living-room, down a couple of steps, there was a bathroom in the adjoining house. A door between me and my neighbour was kept permanently locked, which did not exclude the flow of dialogue from either side.

Anyhow, I was installed in London. I was writing. Apart from the juvenile plays, I started on what became my first published novel (which I went to St Jean-de-Luz to finish, because everybody wrote some of whatever it was abroad). Why I started *Happy Valley* and went to St Jean-de-Luz was because about this time I met the painter Roy de Maistre.

Back in my childhood I had started hearing about Roy, whose brother Etienne (Etty) had married Lottie, my godmother Gertrude Morrice's sister, and become famous training race-horses for Sol Joel the South African millionaire. Listening to the gossips who were unaware of my interest in their theme I had gleaned that Roy, by Australian standards, and in contrast to his brother, was in some way peculiar, an artist who had started by playing the viola in the Sydney Orchestra, developed tuberculosis, turned to painting, gone away, and never come back, or only briefly. The ladies I was listening to, as they snipped at their work in the sewing room, really put their heads together in telling how Roy even used to trim hats.

I had never met him. He became one of my ambitions, a myth I had to see assume flesh. It did not happen for some time, and I can't remember how it did. It must have been in Ebury Street, probably from our own engineering; he had certainly heard about Gertrude Morrice's godson. Anyway, we met; like all important encounters, whether good or bad, it had to happen.

Roy was then living in a groundfloor studio at the back of an Ebury Street house. The first time I went to see him I felt I was embarking on a voyage of discovery. The narrow, white boarding of the studio walls together with white curtains did in fact suggest an actual ship. Through the great windows along one side, a sooty, fog-bound yard became in my eyes a mystic garden. There was an austere minimum of furniture. As for the paintings I hardly dared look in their direction after realising I understood nothing about them.

I was pretty well tongue-tied. I could feel myself behaving like the complete leaden bore, while he tried drawing out of me ideas which might have persuaded me they were worth expressing. He was very patient, more than kindly; Roy's relationships with either sex began as courtships, of which he was often unconscious. I fell in love very quickly.

He was a short, thickset, bald man, more like a banker than an artist. Broody middle-aged women who saw themselves as Frieda Lawrence, or who imagined they had found the guru they were looking for, went down like skittles before the de Maistre charisma. That he didn't oblige them didn't seem to matter. They continued turning up, for painting lessons, bearing food in times of sickness, kneeling at the sickbed hoping for illumination, supporting him financially during his impoverished old age. I was jealous of a lot of it, particularly the randy lady who called him her little bull and sent him the present of a blanket with the message that she liked to think of it lying on top of him.

At the time when we met, his pride had not yet recovered from the defection of a lover. A more superficial character might have used me briefly as a substitute, but after the initial skirmishes, the relationship for which I had been hoping developed instead into a fruitful, lasting friendship. I had no means of foretelling this; at the time I only saw myself as boring, undesirable, gauche. Looking back I think I was probably hoping unconsciously to consummate my love for a father with one who was everything Dick was not. My failure depressed me as much as my failure to communicate with my actual father.

Roy was twenty years older than I. He became what I most needed, an intellectual and aesthetic mentor. He taught me how to look at paintings, to listen to music. He persuaded me to walk in the present instead of lying curled and stationary in that over-upholstered cocoon the past, refuge of so many Australians then and now.

He also taught me to discipline myself as an artist. I had seen him close his door in the faces of casual callers. I thought it a cranky joke. I did not really get the message till I found he did not discriminate. It hurt at first. I saw him as a sour old bastard. Nowadays when I close the door on some importunate, destroyer face I feel as though the spirit of Roy de Maistre is at my elbow supporting me.

His work was what mattered, much as he enjoyed social occasions when the ingredients were sympathetic. If he rationed himself socially, it was not only on account of his work, but because he was often in no position to return hospitality. In spite of a prosperous appearance, shoes polished, suits carefully tended to make them last, there was sometimes little more than a two-shilling piece in a saucer in the kitchen. At a time when the paintings which were his real interest, his ruling passion, appealed only to a few, among them Herbert Read, he used to knock off pot-boilers for a living, bunches of flowers which women like my mother felt they could buy without being laughed at. Herself a regular churchgoer, Ruth once made the remark, 'I couldn't very well live with a crucifixion.' Perhaps she was daunted by the shadow of Popery and her Aunt Lucy's disapproval.

Roy was one of those regrettable Roman Catholics. Brought up a heretic by a Presbyterian mother, he reverted to the faith of his ancestors. He lacked the animosity of so many converts, perhaps considering he had always been essentially a Catholic while growing up nominally beyond the pale. In all the time I knew him he never attempted to proselytise in spite of our strong spiritual attachment. I believe his faith was a genuine one, though it had a worldly tinge; he would have favoured a priest in watered silk rather than a peasant in a grubby cassock.

Alas, he was a snob. For all his dedication to his painting, the Queen Mother could have wooed him away from it. He enjoyed a princess. The Princess Galitzin, who lived a few doors away, appeared at some of his parties. Had I wanted to be introduced I doubt he would have brought us together.

Another distinguished woman played a more important part in his life because at a practical level. Lucy ran errands for him, looked his friends over, and guarded his privacy as closely as he would have wished. Her psychic powers were such that she could hold a sealed envelope and give you the gist of the letter inside. I might have fallen foul of one so gifted, who was at the same time jealous and suspicious. Fortunately we understood and respected each other. Once during Roy's absence in winter, after we had moved to Eccleston Street, I went downstairs late in the night and let myself into the studio, afraid the frozen pipes might have burst. I found Lucy, who had come all the way from her lodging in Chelsea with the same thought in her head. We stood together in an icy blaze of electricity. She was wheezing with bronchitis, snatching distractedly at the high-crowned hat she always wore, stuffed with wads of newspaper for warmth and structure. After exchanging a few words, we went our separate ways, reassured.

Lucy entered Roy's life when he came to London from Sydney in his youth and lodged with her supposed aunt in Chelsea. On her deathbed the aunt confessed that Lucy was her daughter by the second Duke of Wellington, got while she was working as his cook. Though her psychic powers should have warned her, Lucy never recovered from the shock – 'The love of men is unnatural, but oh sir, I'm nuts about the girls!'

The Queen Mother, the frail, mauve-faced Princess Galitzin, Lucy the bastard of unmistakable features and the general's high-crowned hat, all helped tend the sacred flame of nobility, even royalty, whether wholly real, or partly mythic, in Roy's history. It fascinated me then, as it does still, like the thick pancake make-up he wore in old age. His blemishes were superficial compared with my own deeper flaws.

In Eccleston Street, in the de Maistre studio-salon, I met other more or less important people, among them Henry Moore, Graham Sutherland, Francis Bacon, as well as Douglas Cooper, who would start off genial and generous, then turn against those he had taken up. Needless to say I did not dare exchange a word with Moore or Sutherland, and would have avoided Bacon too, had he not been a friend rather than a guest at parties. I got to know Francis when he designed some furniture for my Eccleston Street flat. I like to remember his beautiful pansy-shaped face, sometimes with too much lipstick on it. He opened my eyes to a thing or two. One afternoon at Battersea, crossing the river together by a temporary foot-bridge while the permanent structure was under repair, he became entranced by the abstract graffiti scribbled in pencil on its timbered sides. Alone, I don't expect I would have noticed the effortless convolutions of line he pointed out for me to admire. To discover something as subtle as it was simple made me feel quite elated. In those days Francis was living at the end of Ebury Street, across the Pimlico Road, within a stone's throw of the Mozart-Sackville brothel. He had an old nanny who used to go out shop-lifting whenever they were hard up, and as lover there was an alderman.

I had moved into the comparatively sumptuous Eccleston Street duplex on the strength of an inheritance; there seemed no reason why literature should exclude pleasure and elegance. My cramped cupboard of an Ebury flat, and Roy's studio farther up the street, I see as belonging to the period of the Spanish Civil War and a certain amount of social conscience. I can't remember discussing politics with Roy. If we did, it must have been very superficially. In his dedication to painting he may have wanted to avoid topics which impinged on it, just as, increasingly absorbed in my writing, I would have had to admit I was not interested in politics, while paying lipservice to the fashionable radical views. I had shaken hands with Auden in Cambridge after the first performance of his least successful play. I remember meeting Spender, his gawky, schoolgirl stance, right hand grasping left elbow behind his back. Neither would have remembered me; I was still too indeterminate, the mouse-coloured non-personality derided by Waterall/Waters. In assessment of myself I can see I have never been good at *meeting* people, unless they are thrust on me, and thus become unavoidable, or we find ourselves face to face by not so pure accident. But in the days before the Second World War the mouse only haunted, grateful for the crumbs of mere recognition, while wanting desperately to belong to something, to devour somebody passionately. It put people off. Of course the real reason for my no more than theoretical involvement with the Spanish Civil War was a relationship with somebody 'on the wrong side'. The ranklings of guilt perhaps intensified my sexual passion. That Roy introduced us indirectly seems to me further proof that he had no interest in politics — or was he perhaps involved with the 'wrong side' himself, through Catholicism, monarchy, and snobbism? I am more inclined to think he was playing the role of harassed parent, doing something 'for your own good' — or wanting to be rid of a nuisance child.

However it was, I can't hold it against him. He did more for me at this stage of my tentative intellectual life than any other human being. Except my cousin Betty Withycombe.

I can't remember whether Roy and Betty ever met. Perhaps for an instant in Ebury Street, which led, yes, to an invitation to the studio, where she sat glowering her disapproval. Betty continued hating Roy from a distance, as later on she disliked Manoly from closer up — or for that matter anyone who encroached on her property.

Four sets of three sisters have figured importantly in my life. To a lesser extent Sue's daughters. Distance has separated me from two of them. On the other hand, Fran the youngest, orphaned early, and in some ways close to me in character, chose to return to Australia, where she has shown me in my old age that it is almost impossible to be a satisfactory parent. In my childhood there were three Ebsworth girls, as though three Ebsworth aunts were not enough. Daughters of a family solicitor, the three girls were considerably older than I, but flattered my child's ego by accepting me as their equal. I have not seen them in later life, but memories of the three contributed bits and pieces to the girls in *Voss*. Then there were Chekov's three, who haunted my adolescence to such an extent they became an actual part of it. And finally the Withycombe sisters, who first appeared while I was still a schoolboy at Cheltenham.

Jack Withycombe, the father, was Ruth's first cousin. He had served with the Army in Malaya, but always wanted to be a painter. To earn a living as a civilian he worked for the ordnance survey, helping make the maps Betty and I would consult while walking over most of South-West England. Jack still wanted to be a painter. He had a sensitive talent for landscape in the Constable tradition. While the girls were small children the family had lived in East Bergholt. A quiet, rather inarticulate man with knobby finger joints, Jack was the picture of his own paintings. I always felt at home with him, though we failed to communicate to any extent in words; we were in some ways too alike I expect, which may have been why Betty took to me.

When I first met them the Withycombes were living inside Southampton, above a dirt-track from which the motor-bikes seemed to tear through their living-room. I was a bit over sixteen, I think, and was taken there by Helen Waddell, not the scholar, but a cousin of my father's who had a house outside Winchester, where I used to spend my holidays.

Cousin Helen was an odd, but not so odd one to be put in

charge of a schoolboy with solitary habits and literary tastes. For once Ruth got it right, though no doubt she felt that by bringing us together she was doing us both good. Helen was a 'Havilah' White. After an unhappy childhood with a step-mother who disliked her and shipped her off to a finishing school at Fontainebleau, she had married a curmudgeonly Scottish doctor years older than herself. Arthur cultivated roses, and bred one of his own, 'Mrs Arthur Waddell', in honour of the wife whose mind he formed according to his atheistic, Scottish puritan principles. Helen worshipped her husband-mentor, no doubt from gratitude for his rescuing her when dumped like an unwanted kitten. Her grief on losing him was so intense she never mentioned her husband by name. There were allusions to a 'friend' however, and in every room of her house an unexplained photograph of the forbidding Scottish face. Helen was tiny, perfectly proportioned, except that a mound of auburn hair and a pronounced forehead made her head look over-life-size. Appropriately, since Jane Austen was her literary idol, the face looked out from a framework of little ringlets. Her heroines in history were somewhat unaccountably Emma Hamilton and the Empress Theodora of Byzantium ('poor dear Theodora — just a prostitute,' she used to confide in lowered tones, as though pleading with her listener to accept a disreputable yet beloved relation.) She spent most of her time reclining on a sofa embroidering kneelers for Winchester Cathedral. 'Of course I only do this,' she explained, 'for aesthetic reasons'; and within the confines of his silver frame the dour face of the Scottish atheist seemed to express grudging approval.

How Cousin Helen decided to bring me together with the Withycombes I can't remember. She had a highly developed sense of family, but more likely she was made desperate by the continued presence of a moody teen-age schoolboy. Anyhow, the meeting was arranged; we were summoned to afternoon tea in the flat above the Southampton dirt-track.

Though suffocated at times by life with Cousin Helen in the narrow house on Weeke Hill with its cypresses and flint-walled garden, I was a bit alarmed, so was Helen, to find herself in such a different context — remote from her brodering, and literary devotions at the shrines of Jane Austen, George Meredith,

Victor Hugo, and Gibbon, whose *Decline and Fall*, her bedside book, she must have got through at least ten times before her death. There we were, a strange twittery pair, the Regency marmoset and the green-faced antipodean adolescent, exposed to the dashing Withycombe girls, and Jack and Ellen.

Ellen the mother was one of my loves, expansive, slovenly if the truth is to be told, her hair always coming down, her seams bursting. At the same time she had her aesthetic principles, whether getting a room into shape for an arriving guest, in the books she read, in her approach to life. Never in the money, she had that instinctive taste which can conjure beauty out of the junk shop into a cottage. The meals she served were in the same tradition of conjuring and rightness. I loved everything about Ellen, 'Nin' as they called her, but Ellen she was, and I used the name years afterwards.

Poor neat Helen could hardly compete with overflowing Ellen on that first or any other occasion. She accepted it nobly, a resignation made easier perhaps because her life had really ended with the death of her Scottish doctor. Helen and Ellen always made kind if amused allowances for each other while remaining worlds apart.

As for the girls, on this first occasion I think they no more than accepted the muted schoolboy cousin from Australia as an afternoon's duty. They were all three much older than I, and launched into life. Of the three, Betty the eldest was the most formidable, working for the Oxford University Press since coming down from L.M.H. Nobody could have found a ladder in E. G. Withycombe's blue stocking. I hardly dared look her way.

Joyce the youngest was more self-effacing, silent, cynical. For many years we were unable to talk to each other. I did not really appreciate her worth till after World War II when I was down with pneumonia in Inez Imhof's bomb-scarred house, and Joyce came up from the country through a freezing winter bringing me a billy of fresh brown eggs. Joyce painted. Never a talent which threw off the influence of the Slade, she ended up teaching in loony bins. Much of her life she spent in rundown country cottages, far from a water supply, and with cats for company. She had a complicated marriage, with a husband who came and went, and worse. But as one of the world's masochists she put up with that too. And allowed herself to be coaxed into a communal house in Maida Vale, to cook and slave for Richard and a coven of amateur actresses, all ideals and sexual urge, peace being the alleged theme. It has always surprised me that Joyce with her nice wit could have come at any of that. I remember Betty who deplored Cambridge (with the exception of Kings, because Kings was more Oxford than Cambridge) bursting out with, 'I wonder what happens to them all — such pups!' Joyce quietly suggested, 'They grow into dogs, I expect, and marry bitches.'

At the time when I met the Withycombe sisters, Peggy the second was the most spectacular. She had been out to Capetown, where Ellen's composer brother was director of the Conservatorium. Peggy had mystery, travel, experience on her side; there were rumours of love affairs, one with an important South African. Already accepted as a sculptress (the Luxembourg had bought a carving of an African negro's torso) she might have gone on from there if bearing six children and the activities of a wayward husband had not sapped her physically and morally. As a young woman she had luck, but it turned. She and Joyce were able to make a long stay on Majorca when Peggy sued a bus company after one of their buses had fractured her skull by throwing her against the roof. But this stroke of seeming luck was leading up to the turning point: on Majorca she met her future husband recovering from tuberculosis. Tom was a handsome man with a body, and what suggested a genuinely idealistic mind. He and Peggy became convinced Marxists, their faith strengthened by Tom's professional visits to workers employed by a cigarette factory in Mornington Crescent. As the firm's medical officer he came in contact with earthly hell in the rotten tenements where most of his patients no more than existed in small overcrowded rooms smelling of poverty, sickness, and sewage. I once went with him on his round. It was my first encounter with the last stages of cancer, and my only experience of a human being suffering the torments of that disease in total squalor and the extremes of social misery. That I was not converted to Tom and Peggy's Marxist faith, if only temporarily, goes against me. I was too well entrenched behind my own egotism and my father's allowance to embrace Communism in any guise. For many years I felt no need for a faith either dialectical or mystical, believing as I did in my own brash godhead. As for Tom, his idealism declined through a series of sexual affairs into a moral seediness. Peggy switched from this to that: from Marx to Mao, to the I Ching, even British Liberalism, ending as she doesn't know what. (One of her remarks: 'I've always been fascinated by nuns.')

Jack and Ellen had been so wrapped up in each other their children were not brought up in a religion. It did not prevent Betty on reaching what is known as maturity from having herself baptised, confirmed, and from becoming later a lay member of an Anglican order. Of Peggy's Marxism she was sceptical but tolerant. Her own political bias was inclined towards the radical; she was what I would call a sentimental socialist: the rich can't help doing wrong, the poor are inevitably right.

There was little love lost between Betty and Peggy. A bedazzled antipodean schoolboy, I was not aware at that first meeting as the motor-bikes started hotting up on the dirt-track below, but their attitudes came into full view when the family moved to an Eighteenth Century house which might have been built for them at West End outside Southampton. By then I had evolved from schoolboy into undergraduate. Peggy had just landed Tom, Betty had her principles and a whole clutch of tortuous relationships, Joyce paced the drive in conversation with Hugh Casson. I vaguely floated in and out the oval drawing-room, sometimes providing a catalyst of sorts without my knowing. The drawing-room at West End was slightly reminiscent of the Long Room in Hayley's house at Felpham where I lived through the second summer of my English life, but West End was never parent-ridden, nor did it have any of the frowsty Victorian accretions, the suffocating constriction of that other house, rented only for a season. In Ellen's house the stage was perfectly set. We made our exits to the garden between long curtains in amethyst velvet hanging from a gilded cornice. Whether there was a flawed glass in the West End drawing-room like the one in the Long Room at Felpham I don't remember. Mirrors there certainly were, some of them only created by the self-absorbed minds of those looking for clues and confirmation: Peggy going upstairs and returning with her Garbo hair-do restored, Betty's disapproval of 'Peg' (her principles still allowed her to turn it into a flawed kind of affection), myself less engrossed in actual life than in the more intense literary discoveries encouraged first by my godmother Gertrude Morrice, and now fostered by Betty who had marked me as her protégé. At this stage Peggy, fulfilled by her relationship with Tom, had no need for a tame antipodean cousin, unless as a mirror when she came downstairs flicking back her freshly washed hair. Even then I think she and I unconsciously understood each other better than Betty understood her still passive protégé. As artists Peggy and I functioned intuitively, Betty had the mind of a university intellectual. It was only in middle age that Peggy and I drew close to each other, she in need of a friend and confessor during the collapse of her marriage in New Zealand, myself starving in an Australian post-War intellectual desert. It was then also that Betty's dislike of Peg exploded, transferring itself to me after Peggy had shown her a letter containing a silly piece of interhemisphere gossip. What Peggy had engineered resulted in my not speaking to Betty for years, she hated Manoly anyway, while Peggy became my elective cousin-sister. It was a warped situation, or lopsided triangle, created by three Geminians.

But at West End in the Nineteen-Thirties none of this had happened. We were still Jack and Ellen's family, in spite of Peggy's worldly experience, Betty's principles and intellectuality, and my own adoptive status. We were all united in enjoying one of those frenetic family giggles round the dining table for a guest who did not fit into our tight scheme. On and on we used to go, to the mystification of the guest, even E. G. Withycombe of L.M.H. and the Oxford University Press, compiler of *The Oxford Dictionary of Christian Names* (Miss Betty of the 'black looks not arf' in the words of Edith Eels, a maid) till we subsided in exhaustion or the hiccups. Miss Betty's looks could certainly disconcert, the frowning forehead, eyebrows like black caterpillars, firmed yet fleshy lips. I quote herself on casually handing me a photograph, 'It makes me look like a Jewish headmistress'; yet there were moments when she saw herself as Lady Blessington. Perhaps a little of each, with a dash of George Eliot thrown in, as the flaws in the glass are multiple. She liked to practise charity, but usually failed, as I recognise from my own failures. My most impressive cousin aspired, I believe, to sanctity, but was held back by intellect and 'Miss Betty's black looks not 'arf.'

What do I believe? I am accused of not making it explicit. How to be explicit about a grandeur too overwhelming to express, a daily wrestling match with an opponent whose limbs never become material, a struggle from which the sweat and blood are scattered on the pages of anything the serious writer writes? A belief contained less in what is said than in the silences. In patterns on water. A gust of wind. A flower opening. I hesitate to add a child, because a child can grow into a monster, a destroyer. Am I a destroyer? this face in the glass which has spent a lifetime searching for what it believes, but can never prove to be, the truth. A face consumed by wondering whether truth can be the worst destroyer of all.

God is everywhere they told me. Is he in the bunya-bunya tree? Yes, everywhere. So much of my early life revolved round the bunya halfway up the drive at 'Lulworth' I was almost convinced. Determined on total conviction I asked, 'Does he tear his pants?' I was told not to be silly, which of course it was, since God was in the image of old Mr Bagot at Bowral, his snow-white beard so much purer than those of my photographed grandfathers, which suggested wads of matted musty tobacco. There was also gentle Jesus meek and mild, in sugar pink. If they had presented Him to me lithe and muscular in the form of Deerfoot, my infallible Red Indian hero, I might have been won for Christianity without a struggle. I just could not accept that Victorian oleograph version of the Saviour Ruth and Dick's pretty baa-lambs were expected to adore.

We were taught the prayers we gabbled to Lizzie, sometimes to Mummy when she came in to show us her dress before going to the party. The prayer I composed and which so shocked my father when he came across me intoning it in the summerhouse at the bottom of the garden was more spiritual if they could have realised. But the nightly prayer ritual made our mother feel she was doing her Christian duty by us. Ruth was humane rather than Christian. She sincerely believed in the necessity for doing good. It never occurred to her that she often forced her charity on its object with what almost amounted to physical violence. Her Christianity belonged in a social context: good works, regular church attendance, and a visiting card framed in brass on the end of the pew.

I seem to remember the visiting card as an affectation which blossomed at St John's Darlinghurst after Ruth had lapsed from St James's in the city, possibly because St James's was too far afield, or more probably after Mr Micklem had left.

Mr, or Dr, Micklem was an ascetic, celibate Englishman who conducted a service considered 'very high'. Mr Micklem could have been the reason so many ladies fainted on steamy summer Sunday mornings, and were supported, in some cases even carried out by vergers, and sat on upright chairs in the porch. I was overawed by Mr Micklem's raven notes. He had the head, I realised later in life, of a saint in an Orthodox icon, or one of those Greek or Coptic heads painted on wood in the National Gallery London. I was fascinated by the whole production at St James's: the Greek head, the voice, the ascetic cheekbones and blue chin, the fainting ladies (you waited for the next) and Ruth tiptoeing back, hands folded over her stomach, after drinking the communion wine. I never fainted, but was terrified if ever a lady called Una de Burgh in charge of the choristers approached after the service (she had a club foot) to ask Ruth, 'When are we going to have him for the choir?' I used to pray my voice would crack there and then. But my mother knew enough to protect me from Una.

The communion wine . . . In the asbestos church at Mount Wilson, which Ruth would have liked to burn down if it would, where a draft horse had kicked a hole behind the altar, and Mr Heffernan came one Sunday in every month to mumble from out of his white moustache, Matt playing the harmonium, and Flo singing the hymns in her devoted, wobbly English voice, I went one week-day morning into the cold, dark, deserted vestry, the screen of tree-ferns fretting its asbestos wall, and took a swig out of the bottle. It tasted nasty, like any old portwine heeltaps. I confessed to my would-be incendiary mother. She was shocked, and told me I must never do or say such a dreadful thing again.

So, it all remained a mystery: nasty portwine, fainting, and Mr Micklem.

Actually Dr Micklem was a human being. My mother invited him to Mount Wilson, and told me off to take him for a walk, for the good of my soul and to get him off her hands before lunch. He was tall, not so gaunt as when performing at St James's, and his gravity had a twinkle in it. I was soon in love. I decided to take Mr Micklem to the cave I had never shown anybody, and where I kept a cardboard box full of secrets. The cave was on a track to one of those tedious Australian, would-be tourist attractions called Chinaman's Hat. Hidden in the scrub below the cave was the rudimentary ladder I had made by nailing lengths of sapling together. I led the way up the rickety ladder, the doctor wobbling perilously behind me. When I produced the cardboard box and shared my secrets he was graver than ever. He treated me, not as a child, but a conspirator. As he sat in the shallow cave, his knee-caps almost under his chin, I stood a little to one side and behind. I could look down at his bald patch surrounded by cropped pepper-and-salt, almost a tonsure. I would have loved to touch it. Instead we went back to lunch, stowing the ladder in the scrub above the track to Chinaman's Hat. I felt hot, and finally disenchanted.

What the secrets were in the cardboard box I can't remember in detail, but the usual stuff 'written in blood', and Mr Micklem took a wife, and went home to a provostship, escaping the hungry colonial spinsters, kind hostesses, secrets, the lot.

My religious life flowered again in the hygienic chapel at Cheltenham, a copy, but only that, of the one at my Cambridge college. Cheltenham chapel is most memorable perhaps as an image in Lindsay Anderson's film *If* . . . Half the school sat facing the other half across the chapel's central aisle, giggling, blushing, scowling, blank, the carved stalls at a higher level accommodating masters whose idiosyncrasies we knew, often sickeningly, by heart. As I listened to the boring sermons or whatever was poured out on us, while avoiding the headmaster's blackcurrant eyes, my entertainment was in watching faces washed and ironed for official religious worship. Evangelical hymns were belted out antiphonally. They were another form of entertainment. One of my favourites went, 'Be still my soul . . . ' to the tune of *Finlandia*. Blake's Hymn too, was an inspiration, his Jerusalem slowly fading as we trailed back to the house, to grey mutton and, if in season, a helping of three green gooseberries, said to have been squeezed out of Ma B.'s green-tinged face.

At the appropriate age I was confirmed, my chief reasons being that it was something the others were submitting to, and because it promised a break in the monotony. The padre had served in World War I; he had blue eyes and a strong chin inclined to tremble. He was a man of immense sincerity which sometimes overflowed. It embarrassed schoolboys brought up to consider visible emotion weak and shameful. If Dr Micklem suggested the early days of Eastern Orthodoxy, the school padre was descended from the Eighteenth Century evangelists. He left me cold, except when in kindly conversation he inquired whether I would like to ask any questions, and I felt the shame kindling in me as I rootled round in my head wondering what I could possibly ask. I don't think I thought of anything.

Expectation and even religious fervour mounted as the day approached. The Bishop of Gloucester officiated at the ceremony. We ascended in couples for the laying on of hands. Nervousness forced me to kneel on a step lower than the one my companion chose, which made it difficult for the elderly bishop: he had to reach forward, one arm long, the other short; it gave him a deformed look. Perhaps for that reason I remained unaware of the grace my skull was expecting from his fingertips.

Even so, confirmation infused my surroundings with a numinous glow. I was always taking out my little confirmation book to study the rules. For at least a fortnight I was on the point of willing a miracle. I was spiritually refreshed at the moment of communion and in walking down from the altar with a warm glow inside me, and downcast, shriven eyelids in imitation of my mother.

On the day itself, parents of the candidates for confirmation descended on the school bearing gifts of cameras, wristwatches, and other expensive toys in the best tradition of the materialistic upper middle-class. My own parents in Australia did not apparently know that presents were the order of the day, and I tried for the moment to persuade myself that my material loss was my spiritual gain.

But any spiritual growth quickly succumbed to the barren soil of an English public school. The chapel resumed its true hygienic functions: hearty hymns were belted out, and someone failed to restrain a fart. My spiritual self has always shrivelled in contact with organised religion, whether externalised in that grisly museum Westminster Abbey, the great rococo bed for an operatic courtesan in St Peter's Rome, or the petulant Orthodox communities of Mount Athos. I have come closest to what one always hopes for in Ayia Sophia, Constantinople, alone in the Parthenon on winter afternoons after the Germans had been driven out, in the Friends' Meeting House at Jordans, Bucks, in a garden full of birds, in my own silent room. All of them moments which remain inklings rather than confirmation. The ultimate spiritual union is probably as impossible to achieve as the perfect work of art or the unflawed human relationship. In matters of faith, art, and love I have had to reconcile myself to starting again where I began.

On the day the British declared war on Hitler I was in Portland, Maine, U.S.A. waiting to be picked up by friends I was to stay with at Bridgton. Arriving early by train I spent the afternoon at *The Wizard of Oz*, an American myth more convincing than a real-world situation in which an insignificant young man in an English tweed sports jacket was unwillingly involved. Who in the Portland movie house had heard of his first novel? Who in the world would ever read it? As I saw my literary hopes carried away by the flood of history I was gratefully lulled by *The Wizard*. Throughout the war I seemed to exist on several levels: in the higher reaches a swinging trapeze set in motion by Spengler's predictions which I had been faced with in London as an idler of the Nineteen-Thirties, later my discovery of Dostoevsky's *The Possessed* in a transport plane out of Jules Verne above the jungles and deserts of Africa, farther down there were the ironed-out press and radio announcements, while at the lowest level one clung to Judy Garland's Technicolor rainbow and, as the Stukas flew overhead in the desert, that other voice with the catch in it, assuring us that we'd meet again.

None of it had happened yet: I was the insignificant young man leaving the movie house in Portland when the session ended. My friends, slightly less credible than when I had seen them last, picked me up and drove me through an evening landscape. Set in pine forests beside a lake fringed with holiday cabins, Bridgton was the kind of American township which perceives the reality of disaster only when its own roofs are carried away or sons killed in a war which is theirs, not anybody else's. After a few days bathing in the tepid waters of the lake, and on washing day a feast of Mom's baked beans from the traditional crock, I decided I must return to London. Nobody could understand why. Nor could I explain. A sense of duty always sounds priggish when put into words. Not before a few hectic sexual interludes and vows of devotion – wasn't it wartime, theoretically at least, in the uncommitted but sympathetic United States? I crossed the Atlantic.

I had left London at the beginning of 1939 to try to persuade someone in New York to accept my first published, now best forgotten novel. After being turned down by practically every American publisher, Ben Huebsch, one of the partners and literary barometer of the Viking Press, must have seen something in the glass, for he decided to take *Happy Valley*. He stuck to me during his lifetime though I was always a liability. After his death his faith in me probably dared his successors to throw me out. The son of a Hungarian rabbi, he was one of a breed of noble Jews who, if they no longer manifest the orthodox aspects of their faith, have absorbed its ethic, and are guided by instinctive integrity. True Jews have become as rare as natural Christians. Ben contributed traits to the Jew Himmelfarb in my *Riders in the Chariot*, from the inherited psyche rather than the outer man, for to outward appearances he was a worldly, often cynical character. The year I arrived in New York was the year Carson McCullers made her mark with *The Heart is a Lonely*

Hunter, and those New Yorkers who read reviews, if not the book under review, were agog with the event — 'How could a young girl know etc. . . . ?' Ben's quiet reply to their admiring incredulity was, 'Surely she can read, can't she?'

My pseudo-success in having a first novel published in New York, all that was happening around me in that city, my travels through the American continent, even a meeting with Frieda Lawrence and Brett at Taos, induced a euphoria which persuaded part of me to dismiss events in Europe, till Britain's declaration of war broke in on my light sleep.

Back in London everything had to some extent changed. Evacuation had begun. There were the gasmasks. There was an increase in cynicism, fornication, jokiness at the war's expense, not yet the shortages, anyway for the rich. There were the same old faces, changed, but not all that much, partly by the circumstances in which they were now living, partly by my absence in another world: Sue, and a brother-in-law I had known slightly but now had to get used to, Ruth who had come over with her faithful Mabel after Dick's death, planning to lead the rest of her life in London and at Glyndebourne, Inez Imhof, and her brother Mario Maranta who as a confectioner and pastrycook was soon able to call any tune he liked in rationed London, Ronald Waters, and Roy de Maistre.

It was good to be back, but soon boredom set in during that seemingly endless stalemate, the music-hall phase of the war when we were hanging out our washing on the Siegfried Line. We never stopped complaining. We went to the same restaurants if we could afford them, and gorged ourselves on luxuries. We went to the theatre as we always had. We rustled up rather more intimate parties, and had adventures in the blackout.

I should mention some of Roy's first tenants in the house in Eccleston Street, for they had an influence on my life if not an immediately apparent one. On the groundfloor front and in the basement, a White Russian started an antique business. An ex-admiral and crazy, though in a boring way, he became my Sokolnikov in *The Aunt's Story*. Two of the most difficult literary feats are to make a virtuous woman interesting and a boring man bearable. Whether I succeeded with Sokolnikov I have not had the courage to find out. There he is, anyway, set in aspic.

In between the Russian and myself a couple of intellectual Viennese sisters with wheedling ways had come to roost, adding to my still limited experience of the Central European Jew. The full horror of persecution and genocide had not yet torn our complacency apart. We still flickered with irritation meeting refugees on the landing; our buried anti-semitism flared over some drama of the dust-bins.

The rapid march of events and then the stalemate in the first years of Hitler's War make it difficult to disentangle from memory my own comparatively insignificant advances and retreats: where I lived and when, what I did and why. At least I know I began my novel *The Living and the Dead* before my return to London, at Sandwich, Cape Cod, in the summer of 1939. *Happy Valley* may have shown a certain number of critics and readers that I was an author; now I had to show myself by keeping it up. I imagine this is how most second novels get written. In this case, what should have been a portrait of London the city, finished up as the dissection of a group living in it. The war forced out prematurely a book which should have remained several more years in my head. Because the times were disjointed, my personal life unsatisfactory, and the future totally unpredictable, I have never liked *The Living and the Dead*. Perhaps it should not have been written.

I continued working on the wretched book in London. During this cynical, uneventful, dispiriting phase of life I became increasingly discontented with myself and the novel I was writing. I decided at last to return to New York at least till my age group was called up; perhaps by then the book would have resolved itself and I could leave it behind with my publishers. It would have been impossible to get out of England had I belonged to it, but as an Australian I was given a visa. Having escaped from Australia when I left for Cambridge I never wanted to return to it, yet here I was, making use of my Australian passport, and in recurring cowardly moments I was tempted to remain in the States as urged by some of my friends, including one I believed might be the other half of the permanent relationship I had always been hoping for.

On the surface New York was much the same. Adding to its

superficiality one met the more fortunate refugees from Europe and rich defectors from Britain. The greatest change was in myself. A sense of guilt made me irritable and quarrelsome. Though working by day at what I saw as a dispensable novel, there were the nights to fill, and they usually ended in my trailing from bar to bar, sometimes solitary, or worse, in the company of friends on whom I took out my spleen. Sometimes my companions would have to kick me in the ribs to persuade me to get up out of an eastside gutter. They continued knowing me I can only think because I was a writer and an exotic.

I was living with a doctor from the South. He had an apartment with hardly room for more than one in an old house between Park and Madison in the mid-fifties. We believed we were in love. It was a hopeless affair. In less frenetic circumstances, with less irritable guilt on my part, I doubt it would have worked. That small apartment clamped to a disused lift well where I spent my days finishing *The Living and the Dead* would have stifled any relationship. As a Southerner, J. would grow sentimental over the old negress who did his laundry, and refuse to speak again to a friend who had introduced a man light enough in colour to pass for white. I became suspicious of 'Southern charm'. It seemed to me an odious affectation, as it often is. It is also guaranteed to provoke jealousy in a lover. A number of his friends were former patients he had treated for venereal disease. In the state of nervous tension in which I was living, the thought that half New York must be syphilitic was added to my jealousy, my normal literary despair while welding a novel into its final shape, the suffocating small apartment, the revived momentum of the war in Europe, and guilt for my being in the States at all.

The people I met during these distressing weeks could be divided into two categories: those who congratulated me on being where I was and who urged me to stay put, and the others whose eyes expressed a disguised censure or who broke out in overt and vindictive judgment. I had no doubt which were right. One of my worst evenings was at a dinner party given by Ben Huebsch and his Swedish wife just after the news came through that Hitler had invaded Norway. Half the invited guests called to say they could not face a dinner party; it was more than normal New York wartime hysteria. Of those present I can only remember Elmer Rice and a monumental wife he afterwards left for the actress Betty Field, of that memorable exploratory fingernail in the film *Of Mice and Men*. Rice had been born on the island of Manhattan; at that period you rarely came across such a phenomenon. He talked about New York and theatre. I found him stimulating rather than sympathetic; at least his abrasiveness was an antidote to Southern charm. Ben and his dark silent wife, the Swedish Jewess, did not react openly to what was happening in their minds. In all these weeks of stress there was no hint of censure in Ben's attitude to my continued presence in the States, but I sensed that my decision to return to London a second time drew us closer together.

I finished the novel in June and sailed during late summer. It was an immense if also fearful relief. My relationship with J. rose from the ashes to which it had been reduced. The parting was a painful one. We were convinced that in the words of the song we would 'meet again'. A couple of months after my return to England I received my last letter. I continued writing. In my innocence or egotism I could not believe it had happened the way it does. He must surely have died. But he hadn't. Discovering this by roundabout ways, pain turned to rage and resentment, which festered quietly till I realised our relationship had not been more than a botched detail in my life's complicated design.

The Mersey looked extraordinarily peaceful as we sat waiting for officialdom to pass us for disembarkation. None of the normal Liverpool grey outside, rather the golden tones and glaze, the security of timelessness in a Canaletto, except for a touch of menace in the barrage balloons overhead which Magritte might have added to this otherwise benign, unchangeable world.

London too, looked benign, if faded and resigned. As there was a tenant in my flat I started living in a groundfloor-back bedsitter belonging to Inez Imhof's friend and neighbour Josie Gysi, a vast amiable brown Swiss bear, and her husband, a mosquito without the sting. There were the same old faces in the street and voices on the telephone. I resumed most of my friendships, while ignoring those whose disapproval had reached across the Atlantic. Sue and her husband, he safe inside his diabetic condition, were living in his mother's flat at Dolphin Square, a settlement much favoured by conventional Australian expatriates who felt they needed the protection of their own kind. Disappointed that her hat was without a regimental badge to prove she was the mother of a sacrifice, Ruth returned to Sydney with Mabel.

Somebody, most likely Roy de Maistre, suggested I apply for a commission in Air Force Intelligence while awaiting my call-up. I was snob enough to hope for the commission, while the masochist in me felt I should suffer the worst. Waiting for fate to decide, I put in time working for the Red Cross, tracing missing civilians in Europe, sending messages to relatives and friends. Longing for a message myself, the job had a special appeal. We functioned at St James's Palace, part of which had been turned into discreet offices. Those who worked there were mostly bland English-anonymous. One of them asked me whether I came from the southern states of the U.S.

I had resumed a more or less solitary existence. It should not have been a matter for self-pity because I believe it to be the normal condition of most artists. Not that I didn't enjoy the company of friends, food, drink, books, music, theatre, and often exhilarating sexual encounters. But I expected more, and the fact that I did not find it I blamed wrongly on my homosexual temperament, forced at that period anyway to surround itself with secrecy, rather than on the instinctive need to protect my creative core from intrusion and abuse.

The repression society demands of homosexuals obviously reduces them to some extent as members of that society, but if we can quench our fears the perception gained through our temperament strengthens our hand as man, woman, artist, whichever it may be — or all in one. Homosexual society as such has never had much appeal for me. Those who discuss the homosexual condition with endless hysterical delight as though it had not existed, except in theory, before they discovered their own, have always struck me as colossal bores. So I avoid them, and no doubt I am branded as a closet queen. I see myself not so much a homosexual as a mind possessed by the spirit of man or woman according to actual situations or the characters I become in my writing. This could make what I write sound more cerebral than it is. I don't set myself up as an intellectual. What drives me is sensual, emotional, instinctive. At the same time I like to think creative reason reins me in as I reach the edge of disaster.

I received notice that the time had come for the Armed Services to decide whether I was suitable fodder for the war machine. The medical inspection took place in outer London, I believe it was Harrow, in a dingy hall smelling of unwashed body, predominantly feet. We stripped to the waist and sat waiting on benches: flitches, mounds, occasionally even the statuary of human flesh. The almost universal pallor of our ranks was lit by green-to-yellow tones, with here and there a peppering of black or ginger. I was sitting next to a coal-heaver, his skin a blinding Irish white with damp black wisps of hair straggling on forearms and between the breasts. The day was warm, but most of us shivered awaiting our turn.

At last I was directed to a booth where a plump, elderly doctor was reading through my brief history. The examination started quite impersonally. Till suddenly he took off his spectacles and said, 'Look here, you don't really want to get into it, do you?' I forget the exact words of my reply to this offer of a loophole; I expect it was something like, 'Of course I don't, but feel I have to . . . ' He told me his name, and that of his son, a leading actor who had thrown himself off a liner after finding his lover flirting with a beauty queen. I was touched by the kindly doctor's offer, but at the same time humiliated. While showing him I appreciated his consideration, I confessed I had applied for a commission in Air Force Intelligence, so that any aspirations to heroism on my part could only appear half-heroic. In my heart I knew I should be enduring the worst along with the Irish coal-heaver and other lumps of human flesh in sweaty socks.

Not long after, I was summoned to Kingsway and subjected to an interview guaranteed to scare the daylights out of the brashest aspirant. It was my first experience of higher-ranking

RAF officers, their cold bad manners arising from the fact that they belonged to the junior Service and that most of them had not taken part in actual combat. So they took out their sense of inferiority on the raw amateurs, particularly those serving on the ground, among them the 'brains' as the I.O. was referred to. The notable exceptions to the average run of higher-ranking RAF officers I shall always remember with admiration, just as I try to forget the arrogant boorishness of the others.

After the interview I returned to my job tracing lost relatives with the Red Cross, and in Ebury Street the Gysi bedsitter where I spent my nights. The incredible bombs had begun falling. On the opening night (years after, it is only possible to think of the occasion in theatrical or operatic terms) I had been dining at the Café Royal with John Wyse an actor-director who was planning to put on a play of mine, and did so after the war, perhaps persisting through a bond forged the night the first bombs fell on London, for it was not a very good play.

As a man John was something of a dandy, as an actor a Gielgud manqué. He had flair as a director and should have gone farther than he did if life had not tangled his possibilities. The night of the first bombs he was wearing a dashing fedora when we came out of the restaurant. The scent from his cigar mingled with a stench of burning. The eastern sky was ablaze, fire engines clanging in that direction. It was not yet dark. The west was a cold ice-green as opposed to the Wagnerian glow eastwards, the play of light paradoxical as our world was turned upside down, searchlights swivelling above those tethered sky-cows the passive silver barrage balloons, all around us in the Circus and on the Dilly pavements a sinister pattering of metal rain.

What to do without any specific mission? We started for Ebury Street through the rain of flak which most of us disregarded once we got used to it. But on this night of nights we hunched our shoulders and shuddered under our felt hats. At the corner of Eccleston and Ebury Streets, where it seemed I had spent most of my rational life, we were thrown to the pavement by a nearby explosion. Coming round the corner a cheerful soldier in like predicament tried to share his tin hat with the two fops head to head with him.

We heard afterwards a bomb had hit Victoria Station. At the Gysis', and visiting at the Imhof-Marantas, we found the lodgers gathered in the basements, some of them for greater safety in the coal cellars under the pavement. On our discovering by daylight that bombs had ripped through the pavements, the coal cellars lost their popularity. But where to go? Whole houses disappeared overnight, together with people one had known, and seen the day before. There were the Underground shelters, but . . . After the first shock, and experience of the communal basement-kitchen, all grumbles, gossip, snores, farts, and stale air, or under the bed in my own room, or foregathering on worst nights under the stairs with an actress friend Agnes Lauchlan, another long-time lodger in Ebury Street bedsitters, I developed a fatalistic attitude. I usually sat in my room reading, restoring my nerves from time to time with a swig from a bottle of Calvados I kept in the wardrobe.

I remember reading Chaucer's *Troilus and Cressida*, sufficiently remote while apposite – all those Cressids garlanding the war-torn world. Other remembered reading was Eyre's account of his walk from Adelaide to Esperance. Falling bombs and Eyre's Journal started in me a longing for Australia and some kind of creative urge I could not yet channel or even define.

During the early days of the Blitz I heard from the RAF that I had been accepted for an Intelligence Officer. What it involved I had no idea. I was told where to buy a uniform and to present myself in it, again I think at Harrow, but at a different, more hygienic, if equally impersonal building. My fellow recruits were an odd lot I did not get to know till weeks later during an interminable voyage from Liverpool to the west coast of Africa. None of us had any experience of Air Force life or the mystic rites of this 'Intelligence' whose votaries we had become. For the moment we were given little more than our immediate postings.

Harrow I shall always connect with H. A. Vachell, my experience of schoolboys ruder than those who come from Eton, my two equally horrendous wartime interviews, the first as one of the herded mob, the second as a member of the dubious elect, and an incident I read about years later in a newspaper when a team of Dobermann Pinschers escaped from a halted train and tore into an unsuspecting public going about its daily business.

On getting into uniform I could see from the attitude of friends and glances from strangers in the streets that my stock had increased in value, but instead of feeling encouraged, I was embarrassed, knowing that inside the uniform I was still myself. The ability to love oneself is probably what makes an actor; that it also to some extent makes the novelist was something I discovered only much later. In my youth I found it hard to reconcile life and art; if I did, I think I suspected I was committing an insincerity.

I hated that uniform from the beginning. I hated saluting superior officers, which I would not have done in real life, and in the unreal RAF version of it, I had not been taught to execute the hateful salute with the desired quiver of a rigid hand. If I saw a superior officer coming I always made haste to lose myself.

But these were insignificant worries compared with those to come. I found myself posted to HQ Fighter Command as the Blitz was getting under way.

Bentley Priory, a Gothic mansion on the outskirts of London, concealed a ganglion of nerves which reached out through the British Isles from an underground operations room. Pitched into this, I was expected to understand something I had never been taught. Failure to grasp the rules of the game might mean the loss of the whole war, I persuaded myself as I sat watching its progress in Great Britain as plotted on the ops table during the twenty-four hours. We worked to a roster, but even when off duty, in my billet in Stanmore, or on longer leave, in the small flat once occupied by the Viennese sisters in Eccleston Street, I was still following in my mind the unpredictable tracks plotted on the underground table. The reports one had to write were a nightmare. Worst of all, the telephones echoing back at the novice from John o' Groat's to Land's End. Could I identify an object shaped like a lady's purse found in a field outside Taunton? Far worse, the possibility that Churchill himself might ask for a run-down on the evening's air activity while the WAAF in charge of the Intelligence switchboard was enjoying a long-drawn-out session in the lavatory or at the canteen. Fortunately I never met Winnie voice to voice on the telephone.

84

The mess was full of that cold insolence cultivated by regular officers of the RAF faced with a situation which made them unsure of themselves. My billet in the house of a middle-class couple, the man protected by some kind of necessary civilian job, was of a suffocating cosiness, all blue-satin eiderdown and glaring Maple's furniture. I felt an impostor, while the owners went about on tiptoe, trying to convince themselves that they too, were contributing to the war effort.

On longer leave in London I frequented pubs, got drunk, and walked the streets through a metal confetti. I learned a lot about the whore's mentality, and the variations on her one client, in fact the whole tragi-comedy of sex.

After some weeks at Fighter Command I was posted to the Middle East. I was prepared to think it the result of my incompetence, till boarding the freighter on which we were being shipped out, I realised the contingent was composed solely of my fellow recruits from the opening day at Harrow. We decided the Air Force had simply been asked to provide a batch, and we were it, by grace of some exotic, often useless language, like Russian, literary German, phrase-book Italian, or blood links with the Levant. One clerkly type, with strands of hair coaxed across a bald skull, had fought against the Bolshevik. Another, a taciturn Bulldog Drummond, had grown up speaking Russian in Constantinople and Asia Minor. There was the son of a British vigneron from Oporto. A London tailor was also, rather surprisingly, nephew of the Hamburg harbour-master. One character of Smyrna-British origins had just left off monitoring London-Italian telephone conversations dictionary in hand. Though elderly and bald, D.W. still admired his own beauty. He had a photograph of himself as a young man in fancy dress: powdered wig, quizzing glass, and knee breeches. He liked to point out that his calves had been exceptionally fine, and moreover, still were. There was something mysterious, crypto-, pseudo-, about the job lot of us. Nobody believed I was Australian or that I had written the books nobody had heard of. Perhaps in the early stages, not realising what would be expected of us in 'Middle-East Intelligence', we were trying to create an aura of cloak-and-dagger. Some of us suffered a severe shock when faced with humdrum reality on reaching our destination.

Not that the voyage out wasn't humdrum enough, zigzagging in convoy up to Greenland and down to the Azores. Once or twice the sharks' fins we sighted turned into enemy submarines, but only once or twice. I don't remember feeling unduly disturbed, thanks to a sense of fatality the London Blitz had developed in me. Boredom was the worst enemy on that endless voyage. We grumbled, especially about the food, cowheel and tripe always on the menu, and tinned Christmas pudding. We spent Christmas 1940 at sea. We passed the time speculating on one another's real lives, and cutting one another's hair. I made such a hash of one head of hair during a storm, I had to take the clippers and crop my victim to the bone. He did not hold it against me. A shorn skull seemed to add something to his persona. On making landfall he went native, and would materialise out of the darkness, rolling his French 'r's, a white revenant haunting the black continent.

Calling at Freetown after weeks of monotony and self-searching, we were plunged briefly into a cheerful operetta with chorus of black mammies in floral shifts, their heads dressed in Carmen Miranda bandannas, and male supports like two-dimensional sharks cruising and touting through shanty-town streets. As we sat drenched in sweat, supping our tepid peanut soup, vultures hovered overhead, or perched in anthropomorphous trees, or became transformed by motion into a flight of rusty umbrellas flapping after a funeral procession.

Takoradi on the Gold Coast was the main station in our progress towards the war in the Middle East. We spent a fortnight at a transit mess waiting for the aircraft which would fly us across Africa. The newly arrived British showed up somewhat dazed and pallid against red earth and steamy vegetation of a luscious lettuce-green. In the wardrobes, permanently lit electric light-bulbs helped keep the mildew at bay. Every evening at sunset we went down to bathe in a grey, languid sea, in company with the resident whites and their guarded or overtly hungry female attachments, before the nightly gin session.

3   Dick, Ruth (seated), and Melbourne friend in profile, at Bisley, England, 1911.

4   Paddy at 'Lulworth', Rushcutters Bay, *c*. 1918.

5   Dick and his cuckoo, Mount Wilson, *c*. 1920.

6　Sid Kirk as Digger, First World War.
7　Matt Davies the Yorkshire footman before World War I.
8　Flo Davies the Woodstock girl and housemaid.
9　Suzanne with Lizzie Clark in the garden, 'Withycombe', Mount Wilson, *c.* 1924.

The irritation caused by heat and waiting brought on a falling out with D.W. over my dismissal of the novelettes which gave him pleasure, and tunes like Victor Herbert's *O Sweet Mysteree of Life.* He told me more or less that I shouldn't set up as a critic until I had shown what I was worth. It was always difficult when young to deal with such taunts without sounding priggish. I was thrown in on myself, on what was a mingling of conviction and despair for what I believed I had in me, and the grotesquely insignificant part I was playing in a war of gigantic proportions.

But the big bust-up with D.W. came when he found nail clippings strewn in front of the photograph with powdered wig, quizzing glass, and knee breeches. The boy who looked after his room must have used his clippers. I had to put an end to D.'s ranting. It was I who had borrowed the clippers, and in a moment of mental vagary, or fit of pure slovenliness, forgotten to sweep up.

I only mention the episode of the nail clippings as a detail in the background of trivialities, tantrums, adulteries, service feuds, and wanglings for postings and perquisites, against which a great war is fought. My nail clippings had a faint derisive smell after I came across bodies suspended from burnt-out tanks and corpses melting amongst the saltbush of the Western Desert. In the desert my memory began evoking smells rather than scents: hot Sydney asphalt, tomcat on lantana, the stench of burning hoof in the Adaminaby black-smith's shop.

Australia. It was on the map. I received letters from it while stationed on the frontier Eritrea makes with the Sudan. The sense of duty dinned into me by Lizzie had me writing every week to my mother. Ruth complained of not knowing where I was, ' . . . can't you at least tell me the latitude and longitude . . . ' I thought of her less and less as Mum, whereas Mother sounded too austere. 'Dear Maman . . . ' I began my reply the following week, with no attempt at outwitting the censor, and a bit ashamed of my own preciosity. Whether I explained that her request was a foolish one, I can't remember. It was probably too hot, the dust on my skin turning to mud, the squadron monkey playing monotonously with the telephone bell in the little office where I briefed pilots and wrote reports when they returned from their sorties. In time I received from Ruth the reply to my reply, ' . . . what a clever way of letting me know where you are . . . ' It left me puzzled. I could think of no possible explanation of my own brilliance, until one day exploring the map of the countries we were ridding of Mussolini's rule, I came across Maman, a dot on the Sudanese frontier not far north of Kassala, the dustheap where our squadron was stationed.

In the days of peace our mess had been the headquarters of a District Commissioner, its garden an oasis in the dust and thornbush of the Sudan. A well provided us with bath water, and from it we also drank, or rather, watered down our alcohol. At regular intervals a sombre Arab, an *effendi** from the health department, appeared with a jam jar on a string which he lowered into the well, and on peering at what he drew up, declared the water fit for our consumption.

In the D.C.'s garden at Kassala where a shepherd was allowed to graze his flock/herd at sparrowfart while my pilots were taking off on a strike, I learned to distinguish the sheep from the goats. In my boyhood at Walgett sheep were sheep and goats goats; the Bible did not make sense. Here in the false dawn the mixed mob cropping at the few blades of starved grass with the dew still on it was practically homogeneous. Kids and lambs bleated with the same frustration when the dirty calico bags attached to their mothers' udders denied them access. One morning as the light was increasing, a beast lay down on her side and started moaning. The shepherd whipped from out of her an identifiable lamb, soon wobbling bunting at the mother's udder as she resumed her cropping of the grey grass surrounded by her fellow ewes and does, to the tinkling of my telephone clapper agitated by the surly monkey, a red hussar from the Upper Nile, while the sun rose behind the *gebel*.*

Out of a cleft on the summit of that smooth rock, the Gebel Kassala, a scraggy tree was growing. It was said to be the tree of life, and anyone able to climb the rock and gather a leaf would be assured of immortality. In the shadow of the *gebel* and its version of a recurring myth, clustered a village of mud houses surrounding a dusty square. From opposite sides, rival cafés,

Acropolis and *High Life*, faced each other, their proprietors washed-out singleted Greeks flanked by families of strapping Greco-Sudanese children. From being chained to my telephone I never had the opportunity to explore the village, let alone attempt to climb the *gebel*, but still remember the greyish-white rival Greeks and their khaki children seen from a distance. I am haunted by their ghost faces, their mud village invaded by mechanical insects and those who flew and tended them.

On the Eritrean frontier war was not much more than a game against Italians flying inferior aircraft. None of our pilots was killed; nobody went missing for long. The South Africans possessed a crude exuberance and contempt for anything that did not measure up to the little they had experienced in their short lives in the Union. Feuds flared occasionally between Capey and white. The Afrikaner inclined to keep to themselves and their own language; if they opened out at all, it was with those of British descent who spoke Afrikaans. A restricted stock of constantly recurring surnames gave me the impression that white South Africans belonged to the one incestuous family.

At first the complete foreigner, I was soon enjoying hearty extrovert, totally unreal relationships with several of the pilots. It is always agreeable to discover one is liked by unlikely people. I was touched by their trust. They told me in detail about their wives, children, sexual exploits. From being a diffident young misfit chained to a demanding telephone and an even more demanding, petulant monkey nobody else wanted, I was changed by circumstance into a hen fluffing out wings to accommodate a half-fledged, frequently troubled brood.

Knowing that I was only 'attached' to the squadron did not prevent me feeling jealous when a South African was flown up from the Union to take over from the foreign I.O. As he turned out every bit as diffident and inexperienced as I had been in the beginning, my jealousy was calmed, my vanity appeased when the C.O. asked to keep me till I had trained my replacement. Nicholas and I became friends. A Johannesburg judge in later life, he remembers with some inaccuracy the scurrilous conversations we had in Kassala on nights when the mess bar grew fogged with dark-red dust from the *khamsin** and the monkey screaming drunk after all the heel-taps he had knocked back.

The war in Eritrea was over. A signal came at last posting the squadron to Egypt. During a halt at a railway junction en route, I had the good luck to lose the monkey after an emotional chase through a maze of rollingstock which desert moonlight had turned into sculpture by draining it of all possibility of future movement.

It was one of those moments when you are encouraged to feel you have escaped from what is ordained. I stood shivering in that white light, in the somewhat frosty desert air, while sweating, wheezing, from my chase amongst the extended lines of frozen railway trucks. The monkey had vanished. The war had receded. It had never actually reached the Sudan, Eritrea, Abyssinia in more than insignificant ripples, but its blood and suffering were always present in one's conscious mind, as well as churning and heaving in the depths of the unconscious. The flow of events anywhere was suddenly halted at this remote railway junction. Detached from my past, real life, and with no clue to the future, I was temporarily a free being, my physical self pimpling, my scalp prickling, since I had decided to let myself be shorn for greater comfort a few days before the signal arrived posting us north. My reactions to the loss of the wretched monkey were a mixture of regret and relief. I had done my best by it – or hadn't I? Our break was not unlike the ending of the relationship with several of my human partners. Sceptics may argue that these are fictitious memories, that the moment of arrested time and moonlit sculpture in a desert somewhere between Khartoum and Alexandria is too good to be true. I don't believe so. Anything that has remained so distinct can only have happened.

Again time enveloped me. I found my way back to the train taking us to Egypt. Our so-called 'personnel' were dozing, complaining, exchanging nostalgic mental snapshots of the Union inside the packed compartments, or lounging at the open windows to bargain with the vendors of souvenirs and food. The bronze face of South Africa expressed amused condescension, but had a habit of breaking into open contempt for the perfumed wog in his *ghallabiyeh*.* The night might have snarled up badly if the train had not grappled itself to itself, lurched, grunted, exhaled, and pulled out from the siding, carrying with

it our thick fug from sweaty drill, fruit going bad, and those
smelly little Arab hardboiled eggs. At least on arrival in
Alexandria the South Africans claimed to see a replica of
Durban. They felt justified to some extent in travelling the
length of Africa. Soon, with a gulp, they also discovered
Hitler's War as opposed to the aerobatics of Eritrea and
Abyssinia.

In Egypt we were first stationed at Amriya, taking part in the
defence of the Delta. Intercepting the odd recce over Alexandria
was still no more than an approach to the blood and fatality of
war in earnest. The pilots grew bored, and grumbled worse than
ever about their conditions on developing a taste for Alex the
nights we squeezed into a station wagon, or scrambled on to the
back of a truck, to spend several drunken hours going from
cabaret to cabaret with standard Nineteen-Twenties decor and
whore-artistes, a Levantine riff-raff of Arab, Jewish, Maltese,
Balkan, and Hungarian origins. The most difficult part of
the performance came when I had to round up my pilots for
the journey back. From being a pander I became a conscience.
   Alexandria during those war years must have been at its most
frivolous, its most corrupt. The glitter of its diamonds was
betrayed by its values, which were never more than paste. There
was usually a motive behind its hospitality and woozy kindness:
to marry off a daughter to an English Honourable, or better still
a warrant-officer with solid middle-class civilian prospects, or
simply to fuck the troops, or clinch a swill contract with the
Camp Commandant. Most of us loved this eclectic whore of the
Near East, her pseudo-French and Breetish pretensions, her
Jewish warmth, her Greek loyalty and realism. Silken, boring
Alexandria, pinned between the desert and the sea, with no
outlet but adultery and bridge. Campaigns in Libya, the
Aegean, and the Balkan interior were conducted from her bars,
lapsed secrets converted into organised murder. Fortunes were
made out of tea-dust, rice flour, shit, and human flesh. When
the bombs fell they landed largely on the slums and brothels.
The most fashionable whore-house was wiped out by a direct

hit, and with it a clientele of British staff officers. You wondered how much the enemy knew.

In a vast city such as London you could tell yourself during the Blitz that you were not the target. But in a comparative village like Alexandria, its air full of polyglot whispers, you weren't sure, but felt you had less protection. Desert warfare took you several paces farther into actuality: when the bombers flew over, your cluster of tents and trucks and the slit-trench in which you were crouching were the only reason for their presence in the sky.

In the campaign which culminated in the relief of Tobruk and the advance to Benghazi I was part of an Air Force Intelligence section attached to the Army HQ conducting operations. There were four of us in the Air Force detachment: two technical experts on enemy aircraft, and two others fossicking after general information of the kind contained in the scraps of paper the enemy scatters in a retreat.

My close colleague A., another of those mysteries of Air Force Intelligence acting out a fantasy cloak-and-dagger war to salve the ego and justify a dubious existence, was a stork of an Englishman never parted from monocle and flywhisk, too much a caricature to be true. That he was, I discovered in the Second Libyan Campaign: there was no escaping his anecdotes or his pretensions, 'There, you see – Sanskrit . . . ' as he scribbled a character on a pad. There were vague connections with Russia, the South of France, and the British aristocracy. There were hints of personal estates and villas, never substantial clues to family, marriage, any kind of sexual relationship, unless on a night when we were billeted in the house of a half-crazed elderly Italian Jewess in Cyrene, A. muttered from the depths of an agitated sleeping-bag, 'Marriage . . . a very . . . sticky . . . business . . . .'

I doubt whether any of the rubbish the two of us collected threw much light on the enemy forces' composition and future intentions. Our activities were probably only of importance for the novelist in myself. The technical experts dealt with the crashed aircraft, while A. and I rifled the pockets of the dead for maps, letters, diaries. The Italians in retreat discarded their headgear and their boots, and the ledgers of their frontline brothels. Outside Fort Capuzzo we captured a virgin Parmesan cheese. South of Bardia there were Italians reduced to corpses in a battle with Maoris, the yellow flesh melting like butter into the sand and saltbush. At El Adem there was the burnt-out tank, an incinerated human figure caught by the ankle as he rose in an arc in a futile gesture of escape from his red-hot metal fortress. The corpses washed up by the sea were the worst: fish had been nibbling at them, they stank of watery putrefaction and the dope with which their crashed aircraft had been treated; any documents in their pockets, flight-notes, love letters and snapshots, were sodden and blurred almost out of existence.

I too, was stinking before very long. In contrast, the Army officers, the Mileses and Gileses, the 'old boys', appeared shaven and spicked up even at the moments of worst crisis. They were better equipped than the upstart Air Force, and had a tradition to live up to. We were the larrikins of the desert, particularly a ramshackle detachment like ours, amateurs cast by fate in a classic play. I grew dirtier and dirtier during the campaign, hoping the patina I was cultivating would protect me from the winter cold as I slept in the back of an iron truck wearing every garment I possessed. The two technical experts were not far behind me in slovenliness, only A. kept up appearances, shaving with ice-cold brackish water from a chipped enamel mug, his knees mauve and bony between sand-coloured stockings and shorts in summer drill. A. stalking against a sunset or a sunrise, flicking his whisk at non-existent flies, the monocle clenched into his skull, the thin nose dribbling on to a colourless moustache, had obviously joined the wrong service. He belonged with the 'old boys', who tended to reject him however, when we foregathered at the bar in the mess-tent at night. Had he, perhaps, been to the wrong public school? One never found out for the baroque screen erected between A. and the past, and for that matter the present.

At the moment one realised Rommel had encircled the Third Army during its attempt to relieve Tobruk, the Corps HQ to which we were attached was camped on a stretch of wide-open country between an escarpment and the coast. On a windy, wintry afternoon the next move was planned in a hurry. Bob Angas of South Australia landed on an improvised airstrip beside our stranded travelling circus. A blond pirate, familiar to us as flying postman and source of small luxuries, Angas took off in his obsolescent milk-run bus and flew Brigadier Harding into the fortress of Tobruk. At Corps HQ we hung around unemployed while German infantry on the escarpment kept up a continual sniping. There was no means of avoiding it. As the sand spurted up to the left of me I would move a little to the right, till the same spurt and whine set me in motion back to the left. The snipers must have been poor marksmen or too far distant for effectiveness; I can't remember any casualties. But the experience, a comparatively insignificant one, was unnerving in its continuity and the circumstances in which the game was being played out, on this bleak plain, with no more cover than random tussocks of grass and clumps of saltbush, our dispersed trucks and staff cars, and the officers' mess like a tent left over from a fair which had failed from lack of patronage.

The majors and captains, in signet rings, and battle dress which had managed to preserve a crease, kept up a show of heartiness as they criss-crossed one another awaiting instructions. A tinge of mauve had invaded brick-red complexions. As yellow as a home-made wax candle, my colleague A. stalked back and forth, fly-switch at the ready, under a sky streaked with crimson and soot. We avoided each other. I've often wondered whether A. found me as avoidable as I found him.

My one bright memory of an afternoon of disarray I owe to the mess corporal, a big rosy English clown with whom I had struck up a friendship early on. As the majors and captains got on with their campaigns and country-house reminiscences against the bar on more peaceful evenings, Truscott and I developed a kind of music-hall exchange. Nobody else understood what we were on about; they put it down to a drunken aberration on my part, and the quirk of an eccentric servant too invaluable to be discouraged. It did not occur to any of them that we had recognised and appreciated each other's style. When drink ran out, Truscott would nudge me and produce a can, so that our act might continue flowing. The afternoon of the snipers' party was sadly lacking in alcohol for anybody, except that from time to time Truscott's flushed clown's face would appear round the corner of a tent, the meaty hand beckon, and we would go into our act over what was never quite the last can. If ever I hear the porter in *Macbeth* tumbling down the steps to unlock the gate, his 'Anon, anon, sir . . . ' conjures up Truscott of blessed memory as we awaited our orders that afternoon south of Tobruk.

Towards last light we were told that under cover of darkness the Corps HQ would drive in convoy up a corridor between the minefields and, God willing, join the defenders of Tobruk for whatever remained of the siege. The prospect of action lifted me to some extent out of the doldrums, even with A. for company in our dark zigzag through the minefields. The day before the Air Force attached us to the Army I was 'issued with' a truck, a 15-cwt Dodge, and told I should have to drive it. The son of an Edwardian, and having no mechanical sense myself, I had never been a driver. But I spent an afternoon driving the convulsive Dodge over the uneven ground around the camp, and next morning we set off for the frontier, crossing it somewhere below Salum; there is less to run into in the desert. Now the familiar metal casing of the sturdy truck gave me courage in our drive through the night. It was erratic, rough, often with halts which seemed to stretch for hours longer than those we had travelled. We kept up our spirits biting on lumps of ration chocolate. More inspiriting was the variety of British regional voices posted at intervals in the dark to warn us of a *wadi* which might be our downfall, or a cross-track which could lead us astray. A. was reduced to silence for the most part. If we had any cause to speak, we treated each other to the greatest kindness. At dawn we found ourselves crawling through the ruins of Tobruk.

By that first light our surroundings were of a flower-pot colour. Most of what had stood was reduced to shards. The defenders themselves were dust-coloured against a seascape of violent blue. I felt an interloper, an impostor, in what had been described by the service wits of Cairo and Alexandria as a precinct of 'rum, bum, and broken gramophone records . . . '

The rather shameful conglomeration of the Army Corps HQ was immobilised on a stony hillside. The Air Force detachment became automatically unemployed. There was nothing to do once we had inspected the ruins. The RAF was not too popular with the defenders of Tobruk, and our quartet of amateurs not of high enough rank to warrant the rigid courtesy with which the Army received officers of a junior service. In the circumstances, their own refugees, the majors and captains of Corps HQ, most of them suffering from disguised shock, were slightly reprehensible to the stalwarts of Tobruk, or it may have been that Tobruk, after its pulverising, had become a desirable democracy with only a token division between officers and men. We, the refugees, were restoring the British class distinctions the defenders had been persuaded to shed.

As for me, I was happy. Through my forced redundancy I was able at last to isolate myself from my fellows. I had grown fond of several of them, but it was affection only. I was now free to indulge a passion. Here in this dusty, ragged pocket of the war, I dug myself in, unwanted and unnoticed, and got on with Dickens, my stained Everyman volumes propped on the steering wheel of the Dodge.

I read and read. Just as *The Possessed* had been waiting for a certain moment, the flight in an obsolescent Bombay aircraft over the jungles and deserts of Central Africa, Dickens had been reserved for the war in the Middle East and a key point in particular, the three or four days spent in Tobruk at the end of the siege. Detesting, misunderstanding Dickens when I was a boy, I had suddenly cottoned on to him. As blood flowed, and coagulated in suppurating wounds, as aircraft were brought down in flames and corpses tipped into the lime-pits of Europe, I saw Dickens as the pulse, the intact jugular vein of a life which must continue, regardless of the destructive forces Dickens himself recognised.

After the three or four days of our enforced sojourn in Tobruk its relief came as emotional anti-climax to those of us who had been guests rather than defenders. Free to advance westwards on the enemy's speedy withdrawal, some of us saw our progress as only a half-triumph. The tragic details of an enemy's defeat began to outweigh the comic. There were always the boots and headgear an Italian army sheds in flight, the mounds of excrement in their fortifications, but now in the richer country of the Cyrenaican *gebel* there were the farmhouses of the colonists, identical white cubes in what often became a cubist landscape, everywhere deserted except for a handful of Jews in Cyrene, and in the Vale of Barce with its emerald fields and purple soil a few civilians who had hung on hoping to do business with the conquerors. At the safe spearhead of the advance were the scruffy Air Force fossickers, and those untouchables, the green-tabbed war correspondents. In Barce, somebody pointed out Alan Moorehead. Here I only brushed up against the professional reputation; it was years before I met the man and found him the waxwork so many successful Australians become.

Again in possession of its self-respect, the humiliated Corps HQ reached Benghazi. We pitched camp on a stony rise above the town. The drink had caught up with us again. Seemingly there was nothing to hinder our advance on Tripoli, till on one of those windy afternoons when such things happen, a formation of Stukas came in to strafe. Like the sniping in the desert south of Tobruk, it could have been a flip joke played on the war's least important target. Or was it a portent rather than a game? I decided to take it seriously. As the enemy lost height to deliver their message, I jumped into a slit trench, and when the flurry had subsided, found I must have torn a ligament in an ankle. Farce had taken over once more, at least among the actors playing minor roles.

After a couple of days spent hobbling blue-skinned round the windswept camp, I was given leave. (Any guilt I experienced in parting from the others was dispersed when I heard that our detachment had been withdrawn a few days later and that A. was boring the unwary with his desert exploits, in which I had become his driver.) My leave started auspiciously when I was offered a lift to Alex by a P.A. to the General. The amiable young staff captain, in his ankle-length sheepskin, and the sergeant driving the station wagon, had things far better organised than we knew how in the Air Force: undreamt-of rations, and several primuses hissing away the moment we pitched camp in some deserted farm or blockhouse along the road. I got on famously with my frivolous host, related to an infamous *Tatler* deb. Strolling down the street in Derna, once a colonial pleasure resort, he turned to me and said, 'If you were a woman, I expect we'd have become engaged by now.'

D.W., the Smyrna-British colleague with whom I was shipped out to the west coast of Africa, had given me an introduction to a friend in Alexandria. Baron Charles de Menasce turned out to be the Jewish version of a white rabbit. Though exceedingly rich, his wealth had only protected him in superficial ways. If his nostrils did not scent danger, he would be off into a hysterical laughter more revealing than the sad disbelief lurking in the eyes. Round the corners of his mouth clung that faint webbing which cynicism leaves on those too tender to have faith in others and, worse still, themselves.

The grandfather, photographed in sepia smoking his hookah, had been ennobled by Franz Joseph for the loan of a house during an imperial tour of Egypt. A conventional portrait showed that his son had acquired enough panache to present himself convincingly as the *baron* his credentials proclaimed him. Judging by a companion portrait, the *baronne* must have been their son's undoing, as Proust's mother was his. It couldn't be said that the Baron Charles regretted his undoing; if it hadn't been for his timid nature he would have lived up every moment of the consequences.

Full of grand gestures like gifts of canteens and ambulances to the British armed forces, he was unconscious that some of his minor inspirations were comically mean. On one of my last visits to Alex he kept ringing to tell me he had a present for me; he was so excited about it, we must arrange a meeting so that he could hand it over. When finally the meeting took place he produced a pipe he told me he had given up smoking, with it a tin containing some dry crumbs of Four-Square Tobacco he had not had time to finish before making his moral decision. He was so delighted at the idea of my inheriting his used pipe and smoking the last of his stale tobacco I had to accept with a show of enthusiasm.

There were many signs that this sophisticated millionaire had remained a child disappointed in his hopes of affection. Most of his friends were flibbertigibbets and sycophants. Some of these complained that he did more for me than he ever did for any of them. Certainly he provided me with shelter whenever I came to Alex on leave, and meals if I wanted them. He would take me with him sometimes to the houses of solid Jewish families, to most of whom he was related. He liked to hear my reactions, and enjoyed a giggle. But for the most part, while I was his guest, each of us went his own way. He was seldom at home. I felt oppressed in his silent, shuttered flat amongst the Gobelins, chinoiseries, and *fin de siècle* bibelots as ugly as they were valuable. The best part was turning on the bathroom taps and watching the water gush out in a cloud of steam. A silent *suffragi* drifted in and out of the rooms. Down in the street a *boab*, black as a water-buffalo, guarded the door after a sleepy fashion. The other side of my bedroom shutters grew a flamboyant which poured the light from its crimson blossom through the slats. My occasional, solitary meal in the flat was cooked by Louise, a native of Tarbes, who never stopped complaining that *Monsieur le Baron* did not eat at home, when in fact it gave her unrestricted time for entertaining her Free French *matelot*. Occasionally there was a formal lunch attended by relatives and important members of society, vast indigestible meals, the same dishes on every occasion ('Did you have the duck?' the uninvited would ask. 'And the *riz à l'impératrice?*') The Baron deplored the food he was unable to resist. At night he made a point of not eating – till on coming home he gorged himself on cold duck from the fridge. When he died, luckily before the foreigners were driven out (Charles could not have lived anywhere but Alexandria) we were told by Lena Green, an elderly English Jewess who returned from Egypt to exile in Earls Court, 'You know they found 'im stuffed full of duck.'

His gifts of ambulances and canteens were less important to Charles than what he saw as his war-work: most evenings he used to sit inside the doorway of a club for Other Ranks and mark an exercise book with a stroke for every serviceman who came inside. This complicated, yet naif man was delighted to have found employment. There were also the receptions in one of the minor salons of his flat. Class and rank were unimportant. Still, the company remained select. The Baron drew the line at body smells, and fear of VD always made him wear a French letter.

At one of the Baron's tinkling parties I met someone who gave direction and meaning to what up till now had been a pointless and often desperate existence. 'A Greek, is he?' I asked the Baron. I had never met a Greek before. 'Yes,' he sighed, 'a Greek from here.' In Alexandria I learned that 'from here' was a pejorative term, as I also found out that members of Alexandria's foreign communities were closer to Alexandria than the countries to which officially they belonged.

So it was Charles de Menasce who introduced me to Manoly Lascaris, this small Greek of immense moral strength, who became the central mandala in my life's hitherto messy design. Later on, regretting his gift, the Baron warned me, 'Of course you must realise it won't last. It never does with people like us.' As I write, it has lasted almost forty years, and I dare believe it will outlast the two of us.

War has always been against the marriages it makes and intimate relationships of any kind. Impulsively contracted, they are dissolved by distance, assuaged lust, and general promiscuity. In our case there were the added complications of a secret society which convention forced us to join. What had seemed in youth an excitingly dangerous game, often ended, especially in wartime, in squalor and baroque variations. What I had always aspired to was, simply, truthfulness and trust as far as the human body and fantasies allow, and the security of permanence. So much for aspiration. My jealous, not to say violent nature, might have assured repeated wreckage, if chance or Charles de Menasce had not arranged the meeting with M. in July 1941.

While Dick and Ruth were leading the life of the Ideal Couple and cosseting their three months old child in the Knightsbridge flat, Manoly was born in Cairo. It was August 5th 1912. His parents were still fighting it out and would continue to do so till after the birth of their sixth, and miscarriages besides. Manoly was the middle child in the family born to George Lascaris of Smyrna, and historically Nicaea and Mistra, and Florence Mayhew, a Roman Catholic from Vermont, in her background Mayhew of London, President Coolidge, and Aunt Bessie Rorke who sang in the Chicago Opera chorus. On paper Greek Orthodox, George remained untouched by the spirit. I got to know him in Athens after the German Occupation. A charmer for a short spell, he dropped the charm on seeing you were there to stay, when he returned to being a glum, trivial old man whose selfishness had left him droughtstricken. He tyrannised over Despo, the older sister with whom he was forced to live, and bullied his servants; yet continued to cast a spell over young girls, nurses, anybody who might serve his purpose.

Flo I met in her old age, in Florida, when she had reverted to being an Amercian, and on sufferance a Roman Catholic (till George died she sat at the back during mass and was not allowed the sacrament). In Flo's American-Catholic mind there was always the spectre of six Greek bastards, or so I feel. She abandoned them when Manoly was six. Her sisters-in-law had no respect for the American, and she found them ridiculous, as shown on the occasion when she flung a crimson curtain over one shoulder, stuck a chamberpot on her head, and announced, 'I am the Empress of Byzantium!'

George left shortly after Flo. He stalked through the house, twitching the floor-rugs from under his alarmed children's feet, and left for Chicago with a Romanian mistress. The children were brought up by governesses and two dissimilar aunts.

Most Greek eyes wear an expression of fatality, as though brooding over disasters, personal, historic, and those still in store for them. In family snapshots the Lascaris eyes overflow with this fatality. Close-cropped, and in his school tunic, hands resisting the arms of the chair in which he is seated, Manoly is the Greek archetype. Another such was Elly, the eldest of the Smyrna aunts, whom the more perceptive of the abandoned children took as a model. She had those same troubled luminous eyes aware of the Greek fatality. Her sister, the elegant Despo, was a far more dashing, worldly figure. She was thrown from her horse while distributing comforts at the front during the disastrous war with the Turk, and escaped from the sack of Smyrna in a French destroyer. At the same time Despo liked to see herself as spiritual, nibbling at raw carrots and copying lines from Tagore into a notebook, encouraging artists, kissing the hands of archimandrites, and fainting on cutting into a tomato for the first time in her life during the German Occupation. Elly on the other hand was spiritual by nature, perhaps because she was down-to-earth. Both the sisters remained spinsters, though Despo enjoyed an intellectual flirtation. Elly announced as a girl that she would never marry; men's hairy bodies repelled her. Aunt Elly wore the keys. She made jam and rock-cakes, and dragooned the island girls who became her slaves. She was loved because she was professional and just; her severity was overlooked. She read Goethe every morning till the Germans invaded Greece. During the Occupation she starved herself to death so as not to deprive younger mouths of food.

There were other aunts who appeared from time to time playing lesser roles: the Marchesa, who despised Italians, but married Uncle Nico because a Giustiniani had helped defend Constantinople; Aunt Anna the feminist, who founded an orphanage at New Smyrna, Athens, and told us she would die as soon as she had the satisfaction of seeing the Junta overthrown (she did, on the precise day); Aunt Polymnia, Professor of Greek at the Sorbonne, who gave her life to education, was refused a pension for not having taken French nationality, went mad in consequence, and had to be brought back to Athens, where she spent her last days in the Asylum of the Sacred Girdle.

Like most 'enlightened' Greeks, Polymnia was not a believer — superficially, that is. Even the most sceptical cling to the aesthetics, the history of Orthodoxy, until in a crisis, whether some personal quarrel or national disaster, the blood and tears of faith come pouring out through supposedly healed wounds and opened eyes.

The Greeks have survived through their Orthodox faith, professed or submerged. It is also why an unlikely relationship between an Orthodox Greek and a lapsed Anglican egotist agnostic pantheist occultist existentialist would-be though failed Christian Australian has lasted forty years.

What began in the salon of Baron Charles de Menasce in July 1941 was conducted by necessary subterfuge all through the moves and counter-moves of the war years. We learned to accept separation, to avoid scorn by assuming a mask from which adulterous, even promiscuous heterosexual couples are exempt because they are 'normal'. We perfected the art of writing to each other in code.

One of the duties of the I.O. was to censor the letters the airmen wrote. While part of a somewhat redundant Air Force-Army Liaison Wing stationed at Maaten Bagush during a lull in desert warfare, I became obsessed by my role as censor. It seemed as though the scruffy, crumpled letters left in my tray, together with the letters to and from Manoly, were my only connection with real life. There were the bawdy effusions such as the letter from the Corporal i/c Latrines enclosing for his wife the outline of his erection on a scroll of lavatory paper. On the other hand, there were the doors opening, by degrees, week after week, on the tragedy of relationships falling apart. All the while, knowing that I knew, the writers had to face me. I would have liked to share our condition, but this was never possible. I had to remain the officer, the censor, an anonymous figure, guarding the common good against any excesses of human emotion.

As a result, isolation, desert, repressed sexuality, the voice of Vera Lynn on the Orderly Tent wireless, the letters which might never reach their destination or, if they did, convey an uncommunicative message, preyed on me to the extent that my present circumstances began to coalesce with memories of nights in my Ebury Street bedsitter reading Eyre's Journal as the bombs fell on London. A seed was sown in what had the appearance of barren ground. It germinated years later in a public ward of a Sydney hospital where I had been brought from Castle Hill during one of my most violent asthma attacks. In my half-drugged state the figures began moving in the desert landscape. I could hear snatches of conversation, I became in turn Voss and his anima Laura Trevelyan. On a night of crisis, with the asthma turning to pneumonia, I took hold of the hand of a resident doctor standing by my bed. He withdrew as though he had been burnt. While recovering, though still in hospital, I sketched the skeleton of the book I now knew I would write. It was only after returning to Australia and reading a school textbook that I saw the connection between Voss and Leichhardt. This led to research and my borrowing details of the actual expeditions from the writings of those who found themselves enduring the German's leadership. The real Voss, as opposed to the actual Leichhardt, was a creature of the Egyptian desert, conceived by the perverse side of my nature at a time when all our lives were dominated by that greater German megalomaniac.

Voss's controversial origins led to strife with Leichhardt's academic guardians and confusion amongst the thesis writers. All demanded facts rather than a creative act. In time I was forgiven, Voss canonised, and it became my turn to resent the misappropriation of a vision of flesh, blood, and spirit, for translation according to taste, into a mummy for the museum, or the terms of sentimental costume romance. Half those professing to admire *Voss* did so because they saw no connection between themselves and the Nineteenth Century society portrayed in the novel. As child-adults many Australians grow resentful on being forced to recognise themselves divorced from their dubious antiques, surrounded by the plastic garbage littering their back yards; they shy away from the deep end of the unconscious. So they cannot accept much of what I have written about the century in which we are living, as I turn my back on their gush about *Voss*. (If there is less gush about that other so-called 'historical' novel *A Fringe of Leaves* it is perhaps because they sense in its images and narrative the reasons why we have become what we are today.)

As the British Army was again pushed back towards Egypt after its second triumphant advance across northern Africa, I was stationed at Mex. From this placid Mediterranean gulf, in peacetime a resort for the Alexandrian rich, we were supposedly defending the Delta. Again it was an unwieldy headquarters, this time a stationary one. Several of the officers were stockbrokers (hot stuff in the filter room), Levantines who spoke languages, odds and sods like myself who had some literary

German which, it was thought, might be of use in dealing with technical documents. (How much was expected of that literary German! At the end of the war in Europe I was even posted to Florence to help draft the armistice terms, but after a couple of days succeeded in escaping back to where I was more profitably employed and without false pretence.)

Mex was both pleasant and humiliating, invaluable anyway to the novelist in that he had his fingers on the pulse of nearby Alexandria. As Rommel pushed on towards Alamein, fever mounted in the unreal city. Greek society added nought after nought – as Greeks do – to the number of tanks bearing down on them. Anti-British sentiments increased. The services began organising the evacuation of wives and families. During all these preparations I felt myself united more closely than ever, not only with the person who had become the centre of my world, but with a fragile decadent city suddenly compelled to face itself. Alexandria looked and felt like a blown eggshell.

At this point I was posted to a unit forming for some vaguely defensive purpose, destination undisclosed, should the enemy drive through Egypt. After a couple of unhappy weeks in a transit camp at Helwan, my only occupation the censoring of airmen's letters, I succeeded in getting myself posted back to Mex. The C.O., a martinet whom most of those under him disliked, had taken a fancy to me; perhaps an outsider recognising another. Anyway I returned. Rommel was halted. We settled down to the lull before Alamein.

The boredom and futility of war blanketed all those engaged in it. One remembered that other lull when two armies waited interminably either side of the Siegfried Line. Now half my time was spent in an underground operations room in the dunes above the camp, writing reports on an abstraction of air activity as plotted on the table. It was a very drowsy, horse-and-buggy version of Bentley Priory during the Blitz. Occasionally there were expeditions to crashed aircraft to salvage any documents from the corpses of German pilots. Otherwise I wrote letters, gossiped with my equally bored colleagues, drank the pink tea the airmen brewed, and of course, read.

There was every opportunity to write if the war and my surroundings hadn't drained me of all creative will. I did, however, adapt *The Aspern Papers* as a play when it struck me that the dialogue was asking for a theatre. With very little tinkering, the play was there. After the war, when I hawked it round London, I was told, 'Oh no, not in the least dramatic.' Years later, at Castle Hill, I listened to a radio interview with Michael Redgrave who told how he had realised the dialogue of *The Aspern Papers* would make a play if lifted out. As an actor-manager he had no difficulty in getting his version on the stage, and very good it was, as I knew it would be from my sojourn in the underground ops room at Mex while scribbling away in an exercise book, its pages gritty with sand, the text blurred by sweat.

Otherwise I had dried up. There were the stirrings of what I had it in me to write if there were ever a peace; but that didn't seem likely. So I read. I read the Bible, literally from cover to cover. I read *The Peapickers** and was filled with a longing for Australia, a country I saw through a childhood glow. Its people had tended to reject the Pom during the years spent there between school and university. But I could still grow drunk on visions of its landscape.

I remained an amateur in the Air Force. A sergeant who must have been saving it up, bawled me out for slouching past the flag during a parade I was not taking part in. With scornful lip, he asked me where I had received my training. I couldn't help laughing. I said I hadn't had any. Then I slouched on my way, my pipe in my stocking, my Everyman clenched under my armpit. I must have infuriated that sergeant. I would not have wanted to if he hadn't brought it on himself.

Nor would I have wanted to infuriate my commanding officer by telling him that if he forced me to propose a toast to the Monarch during a mess dinner, I would be doing it against my own convictions. That I was coward enough to propose the toast, even with an ironic twist, distressed me for nights afterwards. In this too, I was not quite the professional, yet not altogether an amateur.

A memory of school holidays spent with Clem and Margaret at Walgett. Scene: the kitchen at 'Barwon Vale'. I had boasted to the cook of a liking for vinegar. What caused me to make this assertion I can't remember; perhaps a taste from a teaspoon in the kitchen at home. My aunt's cook promptly poured me a cupful and stood to watch. I started on it, but at once realised the extent of my disastrous boast. I longed to give in, but the cook seemed to be daring me. So I struggled on and downed my cup of vinegar, to the last gasp and a smarting of tears. Then, like some baited animal, I ran away and hid. I could feel my stomach shrivelling up. I was sure I was going to die – but recovered, as from all the cups of vinegar I have been forced to drink in later life.

After the clash of tanks at Alamein and the British Army's final pursuit of a routed enemy across Egypt, into Libya and beyond, the tragic actuality of war receded for those of us who were left behind, a detritus of uniformed bureaucrats playing out an often black comedy at a headquarters in Alexandria. Even before the break-through, while there were still convoys needing protection and savage dogfights above the desert battlefield, war had become something of an abstraction for the pen-pushers writing reports, contributing to a news rag, a digest of the digests, which probably the squadrons never read. One of my jobs was to deal with the press-liaison officers. Whenever I complained that they had altered the truth, they protested that they had to impose their personalities on the information I had given them. It was the first of my many skirmishes with the press, my first experience of the battle between personality and truth.

By now Manoly had been drafted into the infantry of a Greek Army formed in the Middle East. Jollied along by the rhetoric of exiled politicians, and various members of the incongruous Schleswig-Holstein-Glücksburg royal family foisted on Greece at the end of the Nineteenth Century, the army was moved from Egypt to Palestine, to Lebanon and Syria, at one stage after a mutiny, to the banks of the Euphrates, while awaiting the moment when they could offer their blood to best effect in the common cause.

M. did not make a particularly happy private (signals) until his fluent English got him a job as interpreter. There were the usual jealousies of course; most of the Greeks could not understand his excellent English and despised his educated Greek. There was worse to come when as a second-lieutenant acting as liaison officer between Greeks and British, the British liked him. Nor were his table manners those laid down by the authorities in a Greek officers' mess. Nor did senior officers appreciate an eye which saw through their extra-military activities (not so much the sexual frolics as the bags of sovereigns). By way of compensation he had a likeable and honest C.O.

During this phase of the war I was stationed for just on a year in Palestine, at Haifa. Apart from a few recces, air activity was practically non-existent. But something else was in the air. As the main army moved west across the north of Africa, troops were infiltrated through Lebanon and Syria as part of a feint to suggest our invasion of Europe might come in part through Turkey. It seemed an improbable possibility to those engaged in the move. If nothing else, it extended my Cook's tour of the Middle East. I got as far as Aleppo, in a ramshackle train which then trundled on into Turkey. Years later, travelling with Manoly through Greece towards Constantinople in another seedy, slightly sinister train, I finally understood why the approaches to Turkey have inspired so many corny films.

Aleppo in winter and wartime seemed like the end of the world: a grey dustheap slashed by glassy winds, the grey cubes of houses culminating in the arcades of a labyrinthine *souk*. I put up at the officers' club, a grey villa with signs of former magnificence art deco style. My duties were minimal: visits to advance units, listening to the grouses of those who found themselves with nothing to do in a phoney situation. I had experienced it all in several stagnant backwaters of the Middle East War. My hope is that some of those grey-skinned, shiver-

ing airmen, replicas of myself at various stages, may have
brought away some extravagant souvenir to flare out through
the monochrome of old age in whatever their suburbia – as I
have memories of the *souk*, its pools of saffron, indigo, bronze,
buried in the grey Alep dustheap, and a train journey through
the night, not least the long, expiring pauses en route, at
stations like Homs, with their Arab smells of sesame bread and
hardboiled eggs. I remember too, sporadic dialogues with
Greek privates circulating from compartment to compartment
by way of the running-board while we were in motion.

My year-long anchorage, Palestine, was going through one of
the unhappier periods of its history. It was the era of the Stern
Gang. Violence was not yet directed at the enemy without.
Hatred was exploding within, not only that of the Jew for the
British, the Palestine Police, the Arab, but Jew for Jew: there
was little love lost between those of German, Russian, Polish,
or whatever disparate origin. Lacking Zionist ideals, the
unhappiest were the more sophisticated Jews from Vienna and
Berlin, every second man or woman a 'doctor', who haunted the
cafés of Hadarhacarmel, along with *Modedamen* in leather hats
and oil make-up à la Dietrich and middle-aged dandies forced
out of a culture to which spiritually they still belonged.

For its headquarters the RAF had taken over a monastery
halfway up the Carmel, a mouldy building with painted ceil-
ings through which the wet season dripped relentlessly. The
ceiling of the room where we kept the maps reminded me of the
dining-room of Hayley's neo-Gothic folly at Felpham where a
champagne cork had barely missed hitting one of the corner
putti. The atmosphere would have been depressing enough
without any personal dramas. In my section was an airman, one
of the most reliable characters I have come across, whose marriage was breaking up. I had to follow it step by step in the
letters to his unfaithful wife which he gave me to censor. If I had
been older and wiser, I should have thrown off the official mask
and offered human sympathy. As it was, I didn't. I was shy, and
so was he.

Only certain specialised missions made my presence in Haifa

bearable. Otherwise we led frivolous lives in another of those redundant service units. The officers were billeted in a hotel belonging to a Berlin family who did not value their material gains as much as they resented the affront their Jewishness suffered by our being there. Rationing was strict. So we added to our meals, supplements brought in from the NAAFI, and were always using the wrong dishes, unaware of the significant difference between foods *milchig* and *fleischig*. The contaminated dishes were rushed away and sternly washed, but one suspected they never recovered their desirable *kosher* purity.

Then there was contamination by hunger. Some of the Jewish ladies would sell their souls for a tin of ration bacon, all gelatinous fat and what looked like bootlaces. Hunger is the greatest leveller. After the German Occupation, many hungry Greeks were inclined to value tinned food above virtue. I could not help listening in to a conversation between a couple of airmen while we were travelling together one night on our way to Phaleron in the back of a truck. One of them was full of enthusiasm for the uninhibited Greek women. But his mate came up with a more sardonic judgment, 'I wouldna marry wonn a these − you'd come 'ome and find yer house full a bully.' In really hungry circumstances I expect I would have fallen, along with Shulamit and Eleni. In fact, when I was in London after being demobbed, I remember setting out to take my chocolate ration to my two small nieces, and tearing the parcel open, and devouring the contents on top of the bus as we passed Constitution Hill. On arrival I had nothing to offer but my shame. I kept quiet about that.

What gave point to the year I spent in Haifa during Hitler's War were the sessions I had with German Jews who believed they could give information of interest to British Bomber Command. Some of my informants were mildly deluded, others plain crazy as the result of their experiences. One I talked with was a bus driver, an obviously civilised man who turned out to have been director of the opera house in a German provincial town. He had got it into his head that a factory which used to supply the opera with gloves had gone over to armaments. The reasons for his belief were not very convincing, but we pinpointed his factory, and I offered his information to the service bureaucrats in Jerusalem. Whether the activities of the glove factory were investigated we were never told. It was mostly like this; probably nothing came out of my interrogations beyond insights of my own into the characters I came across, and sometimes friendship. But in one instance, the clues provided by one of my craziest and most persistent informants led to such a fruitful bombing raid that we received a letter of thanks and a senior officer in RAF Intelligence was decorated for his pains. My wild-eyed Jew (I think he worked at a bakery) continued erupting into the monastery's mildewed calm. Whether he was rewarded in the way he hoped I had not heard before I was posted.

Looking back at the year I spent in Palestine I see a highly-coloured, degrading collage: murders committed by the Stern Gang, adulteries between the Allied Forces and their Jewish mistresses, bribery, slander, the case of the drunken WAAF officer whose security passes were found at the foot of the German Hospice wall (at the inquiry she pleaded she must have lost them on getting out of the car to have a pee in the grass beside the road.)

Night after night the cabaret churned into action under the windows of the hotel where we were billeted: a thick Jewish version of

> Leaning on a lamp-post
> At the corner of the street
> To watch a little lady go by —
> Oh, me! Oh, my! . . .

followed by an orange female ourang-outang, a 'star' of the Vienna stage, in her Yiddish patter. Down the Carmel *The Warsaw Concerto* was belted out on a piano as we ate our vegetable stews, almost the only good wartime blow-out in rationed Haifa. By day boatloads of European refugees who had escaped through Romania could be seen coming ashore, their few possessions squeezed into shoddy suitcases. Nightly again, at another club, another orange diva, this time Berlin, seduced us with *You Walked out of a Dream* . . . the dream in which we were all walking – of true love, peace, permanence, an Israel visualised by all the Jews, from the tragic European forced settlers, the bloodiest Stern Gangsters, to the tough new race of *sabras* born in the land of their forebears.

Lying spreadeagle on the hard Palestine soil, amongst the stones and fallen crop of a hillside olive grove, something of the Hebrew archetype entered into me. I had my first inklings of Himmelfarb. These and thoughts of an actual Jew, my friend Huebsch, coalesced when I returned to settle in Australia after the war. Incidents began accumulating round Himmelfarb's still shadowy figure. When Manoly and I bought our small unprofitable farm at Castle Hill, we were described locally as 'foreign Jews speculating in land'. There was also the occasion when I shared a taxi in Sydney, and on arrival at my destination was asked for the full fare. I pointed out that half had already been paid. His belly bursting with beer and indignation, the driver stood on the kerb outside the old Petty's Hotel (which later became a blood bank) and shouted at the top of his voice, 'Go back to Germany! Go back to Germany!' The message lodged. I knew what it was to be a reffo in Australia. Just as I was soothed and fortified by Jewish tradition some years later, standing in *tallith** and *yarmulka** beside Joseph Luvis in Sydney's Great Synagogue. All these elements contributed to the writing of *Riders in the Chariot* during the somewhat bitter years at Castle Hill.

I have to admit to a bitter nature. Any sweetness in it comes from Manoly. Determined not to lose the relationship I had found, I must have hitchhiked on and off over the whole coastal road from Benghazi to Tripoli in the Lebanon, to spend a few hours' leave together. Always in the disguise it was necessary to adopt. Sometimes waiting by the roadside in the desert as the sunset lowered like a bad print in a boarding house parlour, I felt the end was approaching, not only for us, but for all humankind. Then the truck would appear out of the distance, and some cheerful sarge or corporal take me on the next few miles of my journey. Somehow I always arrived wherever I was headed.

Once on leave in Beirut, Manoly by now a second-lieutenant, myself a stationary flight-loot, we tramped the streets from hotel to hotel and finally found a lodging in the glassed-off end of a hall in the house of a Russian widow, a Madame Bongov. When M. was paying the bill she asked him if I was his father. Today, both approaching seventy, it looks more than ever like that. Perhaps it is the price a novelist pays for living so many lives in the one body. Only the nape of M.'s neck looks old: quilted and frosted. Surely the nape is the most vulnerable part of anybody's anatomy? Lucky we can't see our own. Unless a threatened professional beauty, revolving between her dressing-table mirror and the hand-glass, catches a glimpse which she dismisses automatically, and feels saved till it happens again.

Most travellers through the Middle East evoke in their writings a picture of burning sand by day and balmy nights under the stars. Those who follow in their footsteps are not prepared for winter when it approaches: the cutting winds, the blue flesh which come with a desert winter, even in a sophisticated city like Alexandria an inadequate heating system – you would see well-to-do families sitting wrapped in rugs round a table, toasting their feet at the brazier underneath. Less disciplined than the Army, Air Force officers tended to stick out the cold for as long as they could in their miscellaneous summer drill. Mediterranean winters were my undoing: bronchitis turning to pneumonia put me in hospital in Alex, and again on arrival in Athens where I had been posted at the end of the German Occupation, in the last phases of the Civil War. To reach Greece and feel as I did was bitterly frustrating. I lay in bed on the outskirts of Athens in what had once been a fashionable girls' school, and watched the snowflakes of melodrama scud past the windows and vanish at once on making contact with a barren hillside, on its summit a whitewashed chapel, that recurring symbol of Greek masochism. I was happy in the sense that I had reached my Promised Land, for which I had prepared with Greek lessons in Alex with Ourania Theraiou, the study of maps and topographical documents available in an RAF HQ, and endless conversations with Manoly about Greece past and present. Around me in the hospital ward I could feel indifference or hatred, in crocks with civilian ailments like my own, as well as men seriously wounded by *andartes** during the British opposition to the Left in the Civil War. There was no one with whom I could share my enthusiasm for a Greece I had so far experienced only in my imagination; it made me intensely restless. A bossy young English army nurse accused me of being the worst patient she had ever had. Most of the army nurses were tough; they had to be to withstand the conflict between wartime and private life. Some of them came on duty with hangovers from the night before, flourishing one of those oldfashioned hypodermics with a needle like a wand, and drove it into the limb of their choice as though skewering one of their own failures.

For most of my year in Greece I was stationed at the aerodrome at Kalamaki which later became unrecognisable as the civil airport for Athens. I was supposed to be helping train the Hellenic Air Force in operational intelligence, which the Greeks could not believe in. 'Intelligence' meant 'Deuxième Bureau'. If you briefed a pilot for the recce of an island in which a German garrison was still holding out, he was likely to cover the frontier between Bulgaria and Greece on orders from his high command, while our own AOC stamped with rage and blamed Intelligence. Or else the pilot, regardless of his official mission, might take off on a pleasure flight to the island of his birth, and delight his family and friends with a display of aerobatics.

At this period, British Raiding Forces, in collaboration with the Greek Sacred Regiment, to which Lieutenant Lascaris belonged, were mopping up resistance on any of the Aegean Islands still occupied by the Germans. They were received with open arms and open thighs. A party of island women running down to the shore screamed at their liberators, 'Here come the great Greeks to caulk us!' Some of the women wore shorn hair to remind them that they had submitted to other caulking during the Occupation. Along with the more overt invitations, there was the domestic sound of respectable young girls beating octopus seventy times on a rock before offering a stew to those who might propose marriage in return.

Athens was a city of contrasts and conflicts in those winter months after liberation, in the final stages of civil war. The less fanatical, or middle-class supporters of the Left had begun trickling back into the city, to lose themselves in what they hoped in their hearts would be a return to life as they knew it before. There was a final skirmish in the outer suburb of Kephissia, where the actress Miranda, the Maud Gonne of ELAS, rode a white horse in support of freedom. The rebels were quickly dispersed.

The city's face remained pocked with bullet holes, the stucco of formerly great houses and bourgeois apartment blocks chipped and peeling. Inhabitants too poor to deal with the black market were still hungry; they did not fit their clothes, neither their old shabby ones, nor the incompatible garments sent by relatives abroad or dispensed by UNRAA. Yet never since have I seen Athenians of all classes so demonstrably happy.

The weather was capricious: forays of snowflakes alternating with hopeful sunlight and jagged windows of blue. There was a procession through the streets when they brought the icon of the Tinos Panayia to the Metropolis, banners and gilded emblems pagan to western eyes slowly parading, accompanied by the great patriot-prelate Damaskinos, the armed services, and pallid civic personages, some of them reputed collaborators. At the Metropolis the icon was exposed for homage. I joined the queue. A peasant woman overcome by emotion at the prospect of kissing the Panayia fell to the ground in what looked like a fit. She was whisked away. The queue advanced towards the vision of glassed-in diamonds outshining a face as archetypal and unadorned as that of the peasant who had just fainted. Though the glass was wiped after each kiss with a wad of medicated cotton wool, I had been brought up to fear 'germs'. Perhaps I could not exclude the divine ones; anyway I bent and kissed the air this side of the contaminating glass, and shuffled on, full of regret for my hygienic Protestant upbringing. It did not even allow me to carry away a close-up of the miraculous face. My consolation was in recalling that Queen Sophia, after dedicating her pearls to the Tinos Panayia when praying for her husband's recovery, took them back after he was out of danger.

Athens always reacts through its gut. Food however crude was appearing again in the tavernas. The streets were full of heartfelt song, the dusk loaded with perfume of stocks, the comforting smells of roasting chestnuts and *kokkoretsi*.* It was still a village; you could meet a peasant in University Street, a lamb slung across his shoulders like Apollo or Christ. The island of Aiyina was still visible across a blue Saronic Gulf. Rising above the city the Parthenon had not yet begun to look like an archaeological artefact; it suggested pure spirit for this last moment in time before human cattle from the four corners of the earth began shaking its foundations as they trample in herds over the Acropolis.

No doubt these will be interpreted as elitist sentiments in 1981, but on afternoons with winter verging on spring in the aftermath of a terrible war, when apart from a custodian I was the only visible character among the *dramatis personae* who haunt the Acropolis, I saw the Parthenon as the symbol of everything I or any other solitary artist aspired to before we were brought down into the sewage and plastic of the late Twentieth Century. Don't despair however, any of you who have continued reading; it is possible to recycle shit. Could this be my positive message to the Australian optimistic jingle-writers of today?

When I was discharged from hospital I began meeting Athenians during the hours when I was off duty. There was Manoly's youngest sister, Elly, one of the Lascaris family's three solid mandalas, the first being Elly the aunt who brought up the six abandoned children and who starved herself to death after the Germans invaded Greece. Elly the younger, when I first saw her, conjured up certain roses and certain apples. In the Athens markets apples had begun appearing out of Thessaly, a rosy vision to one who had spent several years in the Middle East. Elly's crisp, rosy beauty as a young woman and mother was closely related to all that is admirable, all that is real, in the only country to which she could belong. Her second son Notis

(Epaminondas) was born on the kitchen table during the Occupation. (Elly's grandfather underwent a serious operation in Smyrna, also on the kitchen table, anaesthetised by hammer blows, while his wife stood holding his hand.) Elly was delivered by her husband Elias, of peasant stock from Agrinion, the tobacco country in the west of Greece. In the early days of his dedication to medicine, Elias used to ride a donkey to patients in remote mountain villages. He and Elly married against their families' disapproval. The Lascaris of Byzantium found it hard to take a peasant. The Polymeropouloi of Agrinion resented the Athenian intruder as an affront to local brides. The first child was born in a house where Elly used to coax the fire with a turkey-feather fan, her first maid a prostitute examined daily by Elias for the pox. In time Elias became the one on whom the Lascaris family most depended, director of a leading Athens hospital, a heart specialist while remaining a GP at heart: he will respond to a call in the middle of the night from any of his patients suffering from a bellyache as the result of overstuffing with *stifatho*.* Costas, the Polymeropoulos firstborn, became Professor of Thermodynamic Engineering at Rutgers University, Notis the second, has spent recent years researching Chemical Physics at Göttingen.

In Athens I resumed my Greek lessons. My teacher was Julia Pesmazoglou, an elderly Smyrna spinster who spent her time between two sisters, one married, the other widowed, on opposite sides of the city. Julia gave me my lessons in the widow's house, possibly because there was a marriageable daughter. Mrs G. the widow had been a beauty, 'white – white as a statue, and cold' according to Despo, the Lascaris aunt with elegance and spiritual pretensions, who copied verses from Tagore into a leather-bound notebook.

You would not have guessed at the marble beauty of Mrs G., congested as a turkeycock, kneeling on one arthritic knee to wrest from the back of a lower shelf in her dresser a packet of sugar hoarded since before the war. Most of the Smyrna ladies were hoarders. Remembering the Turk, they never failed to prepare for a siege – unless the spiritual Aunt Despo, who fainted on catching sight of a worm in an apple, or on cutting into a tomato (that smell!) during the Occupation.

Not only sugar, Mrs G. had hoarded gossip, which endeared her to a novelist. Rising to her feet unsteadily, clutching the burst packet of sugar, she returned to the fray, ' . . . their grandmother Cleopatra went to bed with half Smyrna – which explains why the family came out slightly Jewish.'

Mrs G.'s flat had been taken over by a band of *andartes* towards the end of the Civil War. She was given no choice. But, she implied, they were decent men. She, her sister, and her daughter, had enjoyed their company; their lives were enlivened by the presence of such unlikely lodgers.

There was a tune played on a concertina which haunted the streets of the Lykavittos my first Athenian spring, the first the inhabitants were able to enjoy since the liberation. It was an exhausted tune, sad but hopeful. I had taken a room in a house on the upper slopes to escape from the RAF officers' mess when off duty. It was another of those ochreous houses pockmarked by foreign occupation and civil war. There was no lavatory, but I was invited to visit that of an adjoining flat. Rather than inconvenience my neighbour and myself, I used to do it in a cigarette tin and throw the contents out of the window. Apart from this, the room was a great joy to one of a solitary disposition.

As I lay on the crude cotton bedspread on those solitary, watery, spring afternoons, the concertina's melancholy tune, breaking and mending, would meander down through the labyrinthine streets of the Lykavittos expressing everything I felt at the end of those tumultuous years, everything of the Greek fatality, which was also my own, and why I was drawn to Greece from a distance, and one Greek in particular. Advancing, the frail tune was quelled, like faint flurries of rain-laden wind, before taking up again at the phrase where it had left off. Like hope itself. Around the corners of the ochreous, pockmarked houses, and into the future.

It was some time before I caught sight of the musician, a small, insignificant man, who could have been a bit unhinged,

10 Roy de Maistre, London in the 'Thirties.
11 Charles de Menasce the Alexandrian.
12 Cynthia Nolan, 1939.
13 Ruth in the 'Thirties.

14   Ruth in Pitt Street, Sydney, in the 'Thirties.

15   Manoly and P.W. on leave in Beirut during World War II.

like most of us who are creative. Walking slowly, deliberately.
Pressing a squeeze-box as though it were part of his own chest.
From which the vital tune was trickling. Winding through the
streets. In and out my half-sleep as I lay goose-fleshed on the
coarse-textured counterpane. Should I risk overfilling the
cigarette tin? Or could I hold out till the concertina-player had
passed before emptying it out the window?

There were the parties celebrating liberation. To one of these,
farther up the Lykavittos, Elly, Manoly, and I took turns
pushing a child's pram with the ice Elly had made from Morello
cherries. The party was given by a group of gilded young Greek
pilots trained together in Rhodesia, whose age had spared them
operational hazards. The women guests were mostly disappointed wives and hopeful spinsters. The cherry ice and some of
the clangers dropped by Catina, the eldest Lascaris sister, were
the high moments of the party.

Another good gossip and inspired clown, Catina continually
shocked her husband Dimitri by interpreting life in physical
terms and imagery. Surviving the shocks his devoted wife dealt
him, and exile amounting to imprisonment for his political
beliefs, Dimitri Photiades became in old age President of the
Greek Writers. At the time when we met he was editor of an
intellectual magazine too abstract in content to succeed, the
author of a play, more historical than dramatic, in which
Miranda of the white horse played the part of the Empress
Theodora. Dimitri finally made his mark with his histories of
modern Greece written in demotic Greek.

In the days after Hitler's War the Photiades, Dimitri and
Catina, were living on the roof of the building where the
Polymeropouloi, Elias, Elly, and their two children, occupied
a ground-floor flat which also served the doctor as a surgery.
Catina could hardly be described as a housewife. In the kitchen
of her cramped hutch you were likely to find one of several cats
stretched out beside the sink, along with the odd shoe, a clutter
of unwashed dishes, and a few ageing *bourekakia.** Catina
accumulated cats. They pissed on the corners of the furniture;
one of them fell from the parapet into the street below. Dimitri

finally put his foot down, and Catina would take the unreliable lift and feed her cat family in the street.

Dimitri was born into another great Asia Minor family. Perhaps due to his Anatolian background, the sitting position has always come naturally to him. 'Ssh!' his wife would warn. 'He is thinking.' Perhaps as an apology for the family estates, Dimitri became a communist. Twice he was exiled to rocks in the Aegean, where his wife was allowed to visit him, but rarely. There is a snap of Catina standing in front of the wire-netting of a fowl-shed where the Photiades mended the temporary break in what has never been less than a conjugal idyll, if rather a disordered one.

As a result of her trials by Greek history, Catina has become an organiser. She got herself a job at the Hungarian Embassy in Athens. She acquired two country properties, and a flat in the city superior to the cramped, converted wash-house on the roof, but could not help introducing a similar clutter into the flat. Catina is synonymous with clutter. When she retired from the embassy, the Hungarians gave her a dinner-set large enough for a state banquet in recognition of her services. The many pieces remain in their cartons all but blocking the entrance to the lavatory and kitchen. There they will probably continue standing as insurance against the future, while Catina's chatter makes her *kephtedes** blench and reduces the rice or spaghetti on her stove to the porridge one has come to expect.

Variations on Athens Transport: the child's pram with the water-ice in it; during the Occupation, the wheelbarrow in which a husband pushed his wife, already in labour, to the hospital where she was delivered of an enormous baby; the buses, the never quite destructible buses . . .

The Athens buses have always been falling apart; in thirty-five years they would seem to be the same buses. You could fracture your skull during a leap over a pothole, break a limb rounding a corner. These ramshackle, immortal contraptions are murderous if you succeed in emerging from a queue never less than eight abreast, to squeeze on board, pressed together with other human sardines, inhaling petrol fumes, cigarette smoke, body vapours, till acceleration churns the victims into a protesting mass, all livery skins, rasping hair, rancid laughter, invocation of saints and the Panayia.

On our excursions to Maroussi in the hectic months at the end of the war, when the scent of stocks from the surrounding fields still prevailed over petrol fumes, Catina was the great queue-jumper. I can hear her screeching at the top of her voice as she mounts the step, protecting her precious husband with an arm. A father protecting his child turns on her and shouts, 'Anyone can see you aren't a mother!' If only one could have explained that she was the mother of this great child, her husband-baby — and however many cats . . .

Elly and Elias gave a party when the war was over. There were those hopeful spinsters expectantly gathered in the small room. (The optimism of Athenian spinsters at this period could only have been brought on by memories of hunger during the Occupation. They were desperate for a meal ticket. Nothing else could account for some of the marriages they made.) The male guests included a few airmen, and pongo officer friends of Captain Aristo, eldest of the Lascaris brothers. More British than the British from spending a war in their army, the Captain could not leave off singing *A Troopship is leaving Bombay* . . . in close harmony with Miss A. who didn't know English. Jilted by the Captain as a girl after he had carried her off through a convent window, her hopes had been revived by rumours from Alexandria. The Captain was not yet divorced, but without a doubt he meant to be. Miss A., still a young woman with faultless legs and dramatic bust, was biding her time. Always knitting, garments for the family, toys for the children, as well as joining in songs on convivial occasions. She had the voice of a peacock from competing against the winds which harry her native island. The Captain obviously appreciated Miss A.'s charms as she lounged against him, an arm laid along his shoulders in her role of *vivandière*, but on getting his divorce he jilted her a second time; he married a girl of fifteen.

The future has not dawned, however. This evening the

Captain and Miss A. are united in singing with appropriate abandon,

> 'A troopship is leaving Bombay
> Bound for old Blighty's shore,
> Heavily laden with time-expired men,
> As well as cunts signing on . . .'

Aunt Despo had a somewhat exhausted voice even at less confusing moments. Now she professes, 'This Miss A. is so kind – attentive – I don't understand what she expects of me . . .' as the words of the song are belted out,

> ' . . . you'll get no promotion
> This side of the ocean,
> So cheer up, my lads,
> Fuck 'em all . . .'

The walls of the small room had begun contracting – joy, relief, expectation, melting, running, congealing. Through the glass doors the remains of the feast Elly had got together in what were still starvation times had dwindled to skeletons of fish and pools of oiled mayonnaise, still symbols at least, of peace and plenty.

> ' . . . so cheer up, my lads,
> Fuck 'em all!'

Aunt Despo stuck it out. In her old age she was an etiolated beauty of true distinction, who had admired the Duse as an actress, and possibly adopted something of her style. I see Aunt D. in her *saloni*, surrounded by leather-bound volumes in the several languages she spoke, photographs of the Smyrna sisters, the spindly, protesting furniture. When young she had literary aspirations. I was shown the fragment of a novel written in prim English, in which Despo as an aviator was forced down behind the lines during the Asia Minor War.

For all her intellectuality, her spiritual airs, and devotion to high ecclesiastics, Aunt D. had her practical side. During Hitler's War when her friend Iphigenia who lived at Porto Raphti brought her the present of a chicken, she took a room in a hotel so that she wouldn't have to share the chicken with her family.

Anyone who has experienced hunger will remember a destroyer of the spirit even greater than lust. Relatives have fought over a bowl of pap left at the bedside of the dying.

I was almost 'time-expired'. The heavy Athenian summer dragged slowly on its course; if I could have made it drag slower I would have. In Greece autumn happens overnight, its rejuvenating powers more peremptory than those of spring. In its wake the November torrents are liable to quench any kind of enthusiasm. The first week in the new year I was due to leave to be demobbed in England.

At the end, so much of my life seemed to have been lived during twelve months in Greece I hankered after returning. Without having been there, Manoly wanted Australia. I interpreted this at first as a desire to spare his family the incongruous situation of our setting up house together, and having to explain it away. There was something of that. Then by degrees I saw myself as the beachcomber all foreigners become when they settle in Greece — tolerated, but never much more than a joke. So I began tearing myself out of what amounted to a dream, preparing myself to face an Australian reality which Manoly could not have begun to understand. It was his illusion. I suppose I sensed it was better than mine.

When I sailed from Piraeus I was still painfully haunted by the thin trickle of a tune squeezed from the concertina-player's chest as he stumped through the streets winding around the Lykavittos, and the almost solid blast of perfume from stocks in the fields fringing the city. All this is gone by now. Jerry-built apartment blocks stand in the fields where the stocks grew; exhaust fumes from unmuffled cars cannot escape from the labyrinth of Lykavittos. Never were there such victims of progress as contemporary Greeks. Peasants who sold their fields in Thessaly and Thrace live like battery fowls on their steel and concrete balconies or expose themselves to television in the cells behind, in every interior the same box flickering the same message. They tell themselves they are happy. They are prosperous, at least for the time being, stuffed with macaroni, fried potatoes, and barbecued meat. Livery and neurotic. The human contacts of village life are of the past, along with those tough, golden, classic hens scratching freely amongst the dust and stones.

We made Cape Malea, moss-green and forbidding by that winter light, rounding it on a glassy swell, and for my part, a surge of history and emotion. The other 'time-expired' remained unmoved. They were going home to wives, sweethearts, jobs, and they liked to think, other material compensation. A few stood jittering inside their greatcoats, bracing their calves against the motion of the ship and thoughts of wreckage they might find on landing.

We disembarked at Taranto, to entrain (actual term) the following day. The slow, freezing journey up the eastern coast of Italy might have been more depressing if a pongo in my compartment hadn't shared his bottle of slivovitz. The train dawdled, stopped, and meandered on through glacial winds, rubbing up against frost-bitten hillsides. Whenever we stopped women in stockings ending below their knee-caps stood below the windows begging. On the hillside above one frozen town a famous shrine must have been deserted by its saint.

We spent a couple of nights and a day at a barracks on the bleak plain outside Milan. The footfall of Fascism echoed through cold corridors. Each morning a frosted stubble stretched towards a pale sunrise. The shop windows in the city itself were resplendent with inessential goods. Elegant couples strolling through the lit streets at night might never have taken part in the war experienced by those peasant women who came begging at our train windows, the Greeks who had lived on weeds during the Occupation, London pulverised by the Blitz. I never revisit Milan without connecting its arcades and Gothic cathedral, its bourgeois respectability, with the black market and black shirts.

Entrained again, we began climbing higher into Europe, Maggiore in the dawn a blue and gold as intense as the illuminations on a mediaeval missal, Switzerland a landscape of frost-feathers and Christmas baubles. At the more important Swiss towns well-upholstered, charitable ladies walked the platforms offering comforts to the troops. By comparison, France was sour; they had suffered too much, shame included in the suffering.

In London they were trying to pick up the threads, doing what they had always done. But alarming gusts of hysteria would blow round the most stable-looking corners. After several landlords had turned me away (a look at my uniform reminded them of unpaid bills and the havoc created by troops on leave in wartime; one ex-butler type burst into tears, shoved me out, and slammed the door) my former landlady Inez Imhof gave me a room in a bomb-damaged house she was patching up by degrees. Inez was a saint, if among the crazier of them. She could do her fair share of screaming whenever the prevalent post-War hysteria took possession of her nerves, and had every cause when Solomon and Sheba the two schnauzers I was taking to Australia pissed on the bed and tore up the new lino in my room.

There was also Mademoiselle Gans, an elderly spinster originally from Alsace, who made a living running errands for the neighbourhood. Fanged and bent, every child's idea of a witch, Mademoiselle had lived with a prostitute for part of the war. So as not to cramp the whore's style she would spend the working hours of most nights in the waiting room at Victoria Station. One morning on returning to what she had thought of as her home, she found it gone, the prostitute and her client with it. This was when Inez took her in. By the peace Mademoiselle was so far advanced in senility she would leave her possessions and purchases in carrier-bags (as I do today) on every landing in the house, irritating her patroness unmercifully. I once heard Inez screaming from her basement full of cats and ironing, 'You're as much use, Mademoiselle, as a fart in a bottle!'

Mademoiselle Gans bought me the horsemeat I fed to my dogs. After cutting up the horse I would go out to some Soho restaurant and eat a goulash of the same soft sweet flesh. For variety there was the whale favoured by Edith Summerskill. Sometimes I used to go to a Polish restaurant not far from where I was living which laid on one of the best cheap post-War blow-outs. One night a mushroom stew had seduced the patrons almost to a man, myself included. As we were lapping it up, Pole after Pole started leaping from table, falling about, vomiting mushrooms. An ambulance was called. I slunk away, expecting effects. On reaching home I found Mademoiselle Gans, who brewed me a pot of coffee – 'It can't be strong enough.' But apparently it was.

When one of my schnauzers got rid of a worm in the back yard, Mademoiselle was thrilled to discover something of a rarity. 'There was never such a worm,' she shrieked, 'it's forked! I must take it down to the Dumb Friends' League.' What became of her I've often wondered – stooped, practically toothless, myopic, shoes going over at the sides. I lost touch on returning to Australia. Somewhere around the world floats a raftful of memorable souls fate has forced me to abandon. How to grapple this human flotsam to me as I am swirled from hemisphere to hemisphere I have never found out. Those of my White cousins who have remained rooted all their lives under the same proliferating tree are enviable, but a feverish temperament like mine could never have followed their example.

Those who had spent the war in London did not appear so conscious of the graveyard it had become for a revenant like myself dropping in from another world. Sometimes the actual graves of those who had died in times of peace still surrounded the crater where a church had stood. I was less obsessed by the material particulars of death than by its infinity haunting the ruins of once stately houses or an expanse of street from which a comparatively humble terrace had vanished. Friends I failed to trace and my own memories of the Blitz added to the feeling that death still hovered over the city I had known. Coming out one evening on to the steps of the house where I was lodging, I was suddenly faced with that orange glare and automatically heard the fire engines. It took me a moment to realise the whole horrendous thing was not being run through again, and that what I saw was only a normal summer sunset.

Shock incidents like this were partly what decided me to

investigate Australia. Another determining factor was hunger. In London after the war it was almost impossible to feel fed, anyway for a gross character like myself. At the mercy of a ration book and a pair of scissors, you cast about in your mind for something to fill your belly. Even bread was on points, everything devoured long before the cheaper restaurants opened with their often nauseating offerings.

So I decided to return to Australia, at least on a recce, to see what I should find other than the landscapes, which had never left me.

I had been demobbed on the outskirts of London where it had all begun for me, this time not at Harrow, I think it was Uxbridge. The Air Force gave us a suit of clothes which promised to shrink to nothing in the first downpour, a pair of shoes with paper soles, a stiff felt hat, a raincoat, and a cardboard suitcase. (I tried out all these gifts and kept them for some years as exhibits in my private museum.)

More important than these material entitlements of the demobbed man was the gift of intellectual freedom. My creative self, frozen into silence by the war years, began to thaw. I bought a packet of foolscap in Victoria Street and settled down at the table in my Ebury Street bedsitter. I started writing the novel which became *The Aunt's Story*. I can't say it poured on to the paper after the years of drought; it was more like a foreign substance torn out by ugly handfuls. It derived partly from an unpublished early novel *The Sullen Moon*, partly from returning to and working over images of the actual past: the paddocks at Moss Vale, the mysterious house on the crater of Mount Ashby across from my prep school, the Morrices' house 'Browlie' somewhere between Moss Vale and Bowral, Brett's* lantern of a house in the Sangre de Cristo Mountains of New Mexico. In writing this novel I had first to break myself of the habit acquired while compiling factual reports in the Air Force, closer to the practice of *objective* journalism than the pursuit of truth in creative fiction. I grew drunk cultivating a garden of words and sensations which had been waiting years to germinate.

Living in Australia after *The Aunt's Story* had been written, published, on the whole scorned by the pundits, and ignored by the public, I realised how much better pleased Australians would have been with my sifted factual reports. A pragmatic nation, we tend to confuse reality with surfaces. Perhaps this dedication to surface is why we are constantly fooled by the crooks who mostly govern us.

Part II of *The Aunt's Story* I wrote in Alexandria, on the balcony of Manoly's flat, opposite the café with radio churning out non-stop Arab music. Safia Zaghloul was a narrow lane inhabited by artisans, Greek-Jewish, Italian, an Egyptian crook who diddled us over a painting contract, and skeletal Monsieur Olivier the upholsterer, whose daughter had married 'un Anglais' and gone to live in 'Ool', Yorkshire, at the end of the war. There was always a certain emotional tension in Safia Zaghloul, especially when the rabble poured down from Kom-el-Dik protesting against the presence of foreigners in Egypt. Then we all closed and barred our shutters. It was snug enough inside.

We were happy with our plans and our dogs in Safia Zaghloul. Manoly was back at the Moya (the Alexandria Water Board) where he had worked before the war, after leaving the Bank of Athens and the quasi-benign despotism of his Uncle Mario, its president. Everybody warned Manoly against leaving for Australia and God knew what. Hadn't the Australian troops wrecked brothels, held respectable old ladies upside down in the main streets, and driven buggy horses into fashionable *pâtisseries*? But Manoly has always been determined, otherwise we shouldn't have lasted together.

All this while I continued writing Part II of *The Aunt's Story* on the balcony with the marble floor, the pots of basil on the sill behind me, and the bookcase at the other end, as the radio in the café alongside churned out the Arab music.

The noise did not detract from what I was doing. In fact it helped bring me closer to what was happening in Theodora Goodman's confused mind, and to consolidate my own fragmented memories of the iron tables and the stone chips in the garden at the Hostellerie de Ciboure across the river from St

Jean-de-Luz when I was finishing *Happy Valley*, my first published novel.

I could not postpone leaving for Australia. Manoly came to see me off at Port Said. All our partings have been painful; this one might have been more so had it not been leading up to what we believed an inevitable objective.

The liner was carrying a cargo of migrants: British, Greek, Maltese, and the usual Levantine mix. There was a Maltese bruiser with Arab wife tattooed from lower lip to chin. There were the C. family, relatives of our Cypriot tailor, a small, round, bald man who would always, you felt, bounce back. On seeing a variety of shrivelled succulent nailed above the tailor's doorway, I had asked him what it was called, and after thinking it out he told me, 'This plant is called nivver-no-dee.' Now his relatives the C. family were sailing into what they hoped would be a deathless future. There were two little girls, like brown mice, who became constipated on the voyage and caused some anxiety. I went with the parents to the doctor to help explain. The father told me that after treatment one of them discharged what he described as a length of black rubber.

During the voyage in 'Strathmore' I wrote the third part of *The Aunt's Story*, seated on deck with my back to any traffic. It would have looked stranger to the migrants if they had not been heading for a strange world; to any Australian it must have looked unforgivable. At least I finished *The Aunt's Story* and had this manuscript as shield of a kind when I disembarked in what was a familiar and at the same time hostile land.

At this period Australia was very hostile to new arrivals. Those of my contemporaries still around tried not to show they had difficulty making contact with me. Friends of my parents' generation were sceptical, not to say afraid, of this curious hybrid produced by my mother for their inspection. Though they knew there were novelists in the world because their wives patronised libraries, what could possibly become of an Australian male of their class who set out to be a professional author? My father would be spared the shame of it. An Australian chauvinist of the old order, he had died in 1937. My mother had been living in a rented house on Darling Point waiting for the war to end so that she could shake off the dust of a country she professed always to have hated.

I appeased Ruth by staying in her house while on this recce of Sydney. The hump which often comes with age was on her shoulders. (I can now catch sight of it on my own if I glance sideways at a shop window.) She was wrinkled, shrunken, she had lost height, but was as imperious as ever. We started quarrelling almost at once. She was frustrated to find that in assembling the jig-saw of her future with the success she always aimed at (a flat in London, Sue and the grand-daughters, theatres, Glyndebourne, the Chelsea Flower Show, a drizzly climate for her complexion, no longer the threat of snakes) one detail did not fit in. The maker's fretsaw, surely not herself, had cut it out the wrong shape, or else she had been given a piece from another puzzle.

I knew more or less before I arrived that my mother and I could not live in the same hemisphere. I became as determined to stay in Australia as she was firm in her intention never to return to it. Perched on an ebony stand in the centre of her dining table there was a flawless crystal bird contemplating its own reflexion in a pool of water. Ruth and I fought out our worst battles at table, voices muted at first, perhaps out of respect for the perfection of that aloof creature the crystal bird, till carried away by passion we grew recklessly strident, while eating our way through Etty's excellent meals. After spending most of her life cultivating her figure Ruth had taken to tucking in. As she demolished whatever it was, the lines on either side of her mouth worked like those in the face of an elderly, greedy man. Indeed, from being married to a mild husband who needed spurring, she had developed a man's temperament, like many other Australian women. Critics in other parts of the world have accused me of portraying almost exclusively weak men in my novels. This might come from living in a country where women tend to dominate the men, even those convinced of their own virility. The women are certainly more interesting till they set out to prove themselves the equals of their men as intellectuals or bullies, when their femininity and their natural talents wither away.

The familiar furniture and colour schemes were everywhere in evidence in the rented house on Darling Point, but what I had accepted as functional and ours in my youth at 'Lulworth' had become interior decoration. Worse, I realised that furniture does not transfer from its true owners in the past to the new-comers who, hankering after 'antiques', acquire the stuff by purchase. Ruth's concern to get away before she dropped dead in Australia made her possessions seem more ephemeral and less hers. The exhaustion of a move to the other side of the world and my refusal to fall in with her plans were telling on her. She was growing vague. She would leave her keys dangling from the front door-lock. In addition to her dread of snakes, there were other fears on the eve of her departure: a thug had dragged the pearls from the neck of an acquaintance who lived a few doors away.

My poor mother was defenceless. Her son offered no pro-tection. He would have been defenceless too, had it not been for the novel he was typing in the small dark room he had taken over, gratefully, for that purpose. Outside, the westerlies blew almost every afternoon. On New Year's Day a hailstorm smashed several of the windows. In concert with a climate which threatened her omnipotence was the invisible presence of Manoly, whom she disliked without having met. She blamed him for all that was happening to her. Yet when faced with him fifteen years later in London, she doted on him, just as when forced to accept as a permanency the man who had dared marry my sister at the outbreak of war, she adored and depended on him. I am pretty sure she was better satisfied in the end with Geoffrey and Manoly than she was with Patrick. I was not resentful because I was not in love with one I found a disappoint-ing mother – as I was a disappointing son: the kind of writer who made her friends uneasy, a communist, everyone was sure, and other things less mentionable. In her bedridden, blind senility, in her Kensington flat, her accusations were reduced to, 'Manoly and Patrick sit there on the sofa mumbling together so that I can't hear what they're saying.' But on Darling Point, where the westerlies browned the hydrangeas in an afternoon, and hailstones came bouncing into the rooms through broken windows, the only moment of truce in the storms and strewn glass of a not very happy relationship, was a good-night kiss on the landing. By then the make-up had worn off a cheek which suggested white, tremulous kid. The thin lips were firm but pale; it was a dry exchange compared with the wet, emotional kisses Lizzie offered when I was a child.

During this visit to my mother, Lizzie came down from her mountain. On Darling Point she was very much a country figure, hands roughened by grubbing in the soil and searching through scrub and grass for the nests of flighty hens. In spite of her stillborn child, her marriage was happy, I think, or else it was like most lasting marriages, an accommodating arrangement between two sensible human beings. She had taken to half-singing half-whistling to herself. Though we loved each other less emotionally, our attachment was still strong enough for us to give no indication of it in the presence of anyone who might grow resentful. Lizzie made me feel quite the Scot when we were together with those who weren't.

Eliza used to come. Now an elderly woman and house-cleaner married to an invalid, she was still the black-haired brown-skinned girl from Belfast I remembered as nursery-housemaid at 'Lulworth'. Eliza cried whenever she addressed me. I cried back. Two such simple souls, we would have quite a crying match.

Lizzie was scornful of Etty, my mother's cook-general ('housekeeper' is today's more exalted term), because she complained of loneliness living on Darling Point. I don't think Lizzie was ever lonely. She was made for the mountain where she spent the greater part of her life, her husband Sid away all day at the saw-mill. She had her garden, and the *Sydney Morning Herald*, particularly the deaths column.

When we were children and Lizzie still the girl from Carnoustie, she would take us on the tram to visit her family at Dover Road, then a sandy outpost of Sydney. The old people were almost unintelligibly Scots, the father caddymaster at the Royal Sydney Golf Club, his wife a little russet apple of a woman, the palms of her hands almost always white from the used golfballs she had been re-painting by the rolling process, her cat Mysie, a sentimental tabby, rubbing and purring at the hem of her apron. Two of Lizzie's brothers were professional golfers. They made me a set of miniature clubs, but I never had the patience for golf (I used to chop up the turf in exasperation.) There were the sisters: Davina the spinster, who died of some mysterious ailment, probably the unmentionable cancer, Georgina and Barbara, both of whom married. After Barbara had a baby we were allowed to look at the mother and child from the other side of a mosquito net. Their fretful stirring was rather awful. I was glad to get away to the smell of re-painted golf-balls and dolichos on the paling fence.

Dolichos was the great draw at Dover Road. I used to take a case with me to fill with seed. My ambition was to see 'Lulworth' taken over by festoons and canopies of dolichos. Lizzie would empty the case at night and burn the seed in the incinerator.

Walter the middle brother was gassed by the Germans in the Kaiser's War. He was a sickly-looking man, rather hunched, with a yellow complexion which could have come either from the gas or the Spanish Armada. He lived to a reasonable age, but played no part in my life. I remember him first when he and Robert the youngest came to Phillip Street to say good-bye before leaving for the war. Robert was Lizzie's favourite. He can't have been more than eighteen, perhaps younger, at the time of embarking. Ruddy-faced, with a white smile and cropped hair, he had the eyes of those marked for early death, as I recognized later from old snapshots of those who have died in wars and from reading the signs first-hand in the eyes of my pilots during Hitler's War.

Like a dog, I sensed something had happened to Robert when on going into Lizzie's room I found her lying, a heap of crumpled white on the bed. It was most disturbing that Lizzie, who dispersed my fears, should be racked by a grief more immense than any I had yet experienced. I did not know how to comfort her, so I crept away.

Lizzie's hatred of Germans set solid. When we were slightly older she once took us on a walk and had us spit on the gatepost of Resch the brewer's Gothic castle. As an old man I described this incident, jokingly, to the poet Nancy Keesing, who wrote a poem referring to 'the righteous indignation of Patrick White's nurse'. She got her laugh, like some journalist, for telling only half the truth, though the fault was mine. If I had projected Lizzie's figure crumpled on the bed, her grief, my childish impotence and misgivings, the poem might never have been written. What to tell and what to leave out while conveying the truth remains the great question. The actor in us cannot always resist the laugh, just as the journalist has to impress his personality, as I found out during Hitler's War when facts I had given correspondents were falsified in the writing up. I am pretty sure Nancy Keesing if led by a Jewish nurse past Resch's gatepost after Hitler's War would have found it perfectly natural to spit. I did not tell Nancy of that other more culpable episode when I spat in my mother's face in the street. I have recorded it in this book, however, with other details which make and mar a portrait.

As an old woman Ruth still loved dressing up. She had evolved and retained a style from the 'Twenties and 'Thirties, the period when this Edwardian Australian countrywoman emancipated herself. She had declared for the lipstick and cocktail age without ever really belonging to it. Because the cocktail party was a mystic rite, she used to maintain *'The Cocktail Party* is my favourite play.' She did not understand what it was about any more than Dick, dropping in on film after film in old age, understood what he was seeing on the screen. Ruth never liked the moving pictures, said they hurt her eyes, and thought there was something common about them.

Brandy crustas and anointing her mouth remained her tribute to modernity; hats were her greatest delight. As the 'Thirties were ending in London, Roy de Maistre asked to paint her portrait. It was really the portrait of a hat. Some of his portraits were abominable; this succeeded, worthy of Ingres, thanks to the hat.

When Ruth gave her farewell party in Sydney before leaving Australia for ever, and invited three hundred friends to drinks at the Royal Sydney Golf Club, I asked, 'Have you really three hundred friends? Some, surely, must be acquaintances', she seemed surprised.

They came, an incompatible lot, including husbands and wives who had parted in dudgeon, now assembled with more recent spouses, glaring, or smiling treacly smiles. Suggesting to my mother afterwards that she might have given a more tactful party, she answered, 'This is what we do in Australia.'

A day or two later, she took off and left us to it. Perhaps she understood the indiscriminate, Australian upper-class party. I hope England understood her better than she understood Australia and me. But I doubt it.

Manoly wanted to live in Australia without having experienced it. Aside from this, what really persuaded me to return after my brief and for the most part disheartening recce was the unexpected art world I discovered. Writers still lived concealed like borer in the trunk of a shrub, but painters were beginning to emerge. Though I have revised my opinion of some of the artists who excited me then, their vitality and colour seduced one who had been starved of any art other than the printed classics during the barren war years. In the same way a flaunting musical like *Oklahoma* bounded across the footlights and seduced a drab, decimated London with its primary colours, primary emotions. All of this belonged to that period.

I met and became a friend and admirer of the painter Dobell. His work was at its peak, though the flaws were beginning to show. A simple provincial boy gifted with the lightning flashes of homosexual perception, he had already accepted that fatal courtship by a society which sees artefacts as investments and artists as poodles to perform at fashionable dinner parties. He was carried away. Later I heard he described me as an 'intellectual snob', while I saw him as selling out on art. It did not detract from his material value. As he grew more facile, his paint more slippery, his stock soared. He became the investment for doctors and lawyers.

To take the sting out of my disillusionment and our mutual disapproval, I like to remember our last meeting. It was at a vice-regal reception in the days before I had become a convinced republican. Bill and I had flopped down on a small sofa in the centre of the room under a great chandelier, and were carrying on the most scurrilous conversation, belting out obscenities as hard as we could. I believe the chandelier tinkled a bit, but the crowd around us seemed unaware of what we were shouting. It was a good end to our relationship, the equivalent of some of his more honest paintings, like that housemaid leaning out of a Pimlico window, the *Sleeping Cypriot*, and his *Billy Budd*.

Dobell's Sleeping Cypriot, my Awakening Alexandrian . . .

As he stepped off the flyingboat at Rose Bay, he looked very small. I was wearing my Akubra* (it was the age of felt, serge, and sweat.) Understandably he was somewhat nervous of the future after touching down at Darwin and Bowen. I was the more nervous of the two and my responsibility for the island-continent to which I had brought the one I most respected and loved. In the same way he has felt responsible for the atrocities and vulgarities of Greece each time we go there. We have learnt to accept each other's anomalous responsibilities, more or less, which the normal expatriate never has to endure, even anomalously.

Just then, what mattered was that we had brought off the seemingly impossible and were standing together on the pier at Rose Bay.

I had booked a room at Petty's Hotel. They kept a good old-fashioned table: duck, roasts of beef and mutton with the inevitable heavy, floured gravy, and standard grills for Australian males and their wives or doxies (some of the wives at that period made a point of referring to themselves as 'mates'.) The bedrooms were stuffy – fluff on the carpets, and a smell of cigarette ash. The more Manoly saw of Sydney the less impressed he became, though grateful to have escaped from Egypt where persecution of its Europeans had intensified. Here we had each other for support, he an expelled Greek, I not so much an Australian as a fake Pom and writer nobody had heard about, posing as a member of my own family.

We had few friends in the beginning, and apart from Clem and Margaret Withycombe and two distant White cousins, Patricia and Morwenna, I had never kept up with relations. The proliferating Whites had always been suspicious of exotics and anybody who might be described as 'clever'. Pat was one of the few Whites who deigned to know me before I got the Nobel Prize. I have always valued her friendship and envied her the equanimity she seems to have achieved in old age with her friend Lyn Garton.

What Manoly and I needed most was somewhere to live and soon, as we had four schnauzers coming out of quarantine. One weatherboard cottage we were offered out at Kellyville set in my memory as a classic Australian image, with its dry-rotted veranda boards, the smell of mutton fat and sick lino inside, and a fig tree growing out of sinkwater. In such a house the O'Dowds would have lived. It is the house Mag Bosanquet connects instinctively with her lover Terry Legge. Eureka Steel lived there with Cuth Spurgeon in the screenplay *Last Words*. It is the house to which Manoly and I return in some of the nightmares of my old age.

In reality we passed this one up and settled for a suburban villa on six acres of paddock facing Showground Road at Castle Hill. It has been referred to as a 'lovely old home' and a 'rundown farmhouse', when it was never more than that sub-urban cottage in painted brick. What persuaded us to buy the place was the fact that there were pigsties which could easily be converted into kennels. We bought the house walk-in walk-out. We lived with the furniture for years, learning about those who chose it, till we could afford what pleased us better.

For a few weeks we were received into the bosom of the family who had sold us the house. In their view it was to soften the blow for two helpless men about to do for themselves. This interlude was illuminating too. Manoly was asked by one who apparently visualised mud hovels, 'Do they have big houses like this in Greece?' They were worthy, often amusing, materialistic Australians. They moved from house to house, giving the one they were leaving a thin coat of paint, we discovered later. When they moved on to somewhere up the coast we were relieved, if also alarmed to be left to our fate.

At least we inherited Mr Patchkoe, a stooped pensioner of solitary habits and Russian extraction, who lived in a filthy room at the back, sleeping under bags. His former employers told us that was what he liked. Certainly in our very brief relationship he never complained. We fed him. He liked nothing better than a boiled rabbit, and a porridge made from farmyard bran: 'it makes me guts rattle.' He left us to go into hospital, and died soon after, a wealthy man it turned out, the owner of blocks of flats in a Sydney suburb. His niece rang us to tell us of his death. He had asked her to do so because, she said, he liked us — which was about the most heartening news we had in our early life at Castle Hill.

Living on what the dentist-owner had described as a farm we felt we had to get into farming. There were two cows. Painfully I dragged the milk out of them twice a day, separating it on frosty mornings or in fly-ridden summer heat, washing the greasy separator parts, at night churning the cream into butter in the kitchen. When we found there was a market for thick cream (illicit as we had no health licence) we gave up butter and quietly sold the cream — our only product to yield a profit.

We started growing vegetables, peas which the frost bit, cabbage which caterpillars devoured, cauliflower with loose curd, broccoli which shot into seed, eggplant nobody in those days wanted. We also grew flowers. There was always a glut in whatever we had. My one poetic memory of our venture in commercial flower-growing is of a sheet of moonlit shastas which nobody bought even when we practically gave them away from buckets at the roadside. We were of course the rankest amateurs, the traditional no-hopers. Those who were alive to the stoniness of life on a small farm stood around, lowering wrinkled eyelids, lips curled sardonically, waiting for us to be driven out by failure. We weren't driven out. We lasted eighteen years, not as farmers admittedly, but as stubborn human beings.

I believe the relationships we formed eventually with neighbours grew out of mutual affection and respect, after they had recovered from our foreignness and 'impostures'. As they began relenting we heard Manoly was thought to be 'some kind of black prince'. Royalty has always appealed to democratic Australians; that they favoured its black side was something of a surprise.

Anyway, the blacks were moving in. Down the road a Sicilian family had bought a farm and were growing vegetables — professionally. Our opposite neighbour who raised fowls and ducks, and sold flowers as we did from buckets at the roadside, became the voice of doom. 'They're black,' she muttered, 'they're black!' She had forgotten Manoly on taking to him. A few months later, gossiping with our Mrs Poulter behind the privet which gave her veranda partial protection, we looked out and saw the Sicilian *signora* advancing up the road. Mrs Poulter waved. The *signora* waved. Smiles were exchanged between black and freckled. Mrs Poulter sighed turning to me, 'She's a lovely lady, she's not black at all.'

So we were established at last. We were called in from time to time to encourage Bill, who suffered from piles, to submit to the knife. But although accepting us as decent enough, he did not trust us to that extent.

In spite of so much that was heartening in the way of humble human contacts, trees growing, shrubs flowering, and the books I wrote chiefly as an escape, I expect, into a more vital world, I hated the years we spent at Castle Hill. It was too oppressive. In that clay hollow, freezing in winter, breathless in summer, amongst the heavy ergot-bearing paspalum, hassocks of Cape weed, rusty rye grass, Patterson's curse, I was constantly ill with asthma. We seemed almost waist-deep in weed. While rotary hoeing our wax-infested citrus orchard, Manoly wrecked his back for ever. We were this pair of amateur actors, miscast through our own determination, or pig-headedness. It was worse for Manoly in that he had attached himself to a prickly character who protested against his fate by throwing saucepans of Irish stew out the kitchen window, cursing, and getting drunk. But we were not driven out by our failures any more than Australian critics in those early days succeeded in killing the creative necessity of one they saw as an intruder, a breaker of rules, a threat to the tradition of Australian literature.

Throughout these years I grappled with the resistant novels I had inside me. From actual experience of a calf twisted in a cow's womb, I now see it as much like that. The earlier part of *The Tree of Man* I wrote at the kitchen table during sleepless nights when spasms of asthma prevented me lying down. Much of *Voss* was written in bed, and after it *Riders in the Chariot*.

It was not all asthma. There were nights of exquisite cool when the southerly arrived from Sydney in the clapping leaves of the camphor laurel which wistaria hadn't yet succeeded in choking. On such nights the mown grass prickled against naked flesh as we lay this side of the sheet of moonlit shastas.

But never for long. I would be standing on the brown-and-pink Wilton carpet, hanging on to the foot of the silky-oak bedstead acquired with the house, my lungs retching after almost non-existent air.

On a memorable night, lying on this same bed, after a session at my desk and too much cheap sherry, I dreamed a conversation with the Devil. I was awakened by the telephone. A voice in extraordinary accord with my dream announced, 'If you care to pursue it, I could lead you much deeper into Jewish mysticism than you've already been.' This was after I had written and published *Riders in the Chariot*, for which I had done some research into the Hassidic mystics. I declined the Devil's offer because I was on with something else. Not long after, a Jewish painter identified the telephone voice as that of his father, who had fallen on his head from the back of a truck escaping from Poland during the troubles. I have often regretted not accepting the brain-damaged Jewish mystic's offer, but this is the price one pays in leading the life of a novelist, part dragonfly, part shark, always at the mercy of the future and his own peremptory depths.

During these years we depended, and very gratefully, on our Jewish migrant friends; they were our link with European culture, music in particular. Through Fritz Krieger, an Austro-Hungarian business man, and his Hungarian wife Ile, I started my Mahler-Bruckner phase. Through them I came to Bartok also, though he was a composer they detested. My relations with Fritz became strained through my passion for Bach. He accused me of being tone-deaf. Then, Mozart: 'a stream of lemonade'. By the time I recognised Mozart as perhaps the greatest composer of them all, we had broken with the Kriegers. The possessiveness of Jewish friends can become suffocating in the end. Like Jewish children who escape from the bosom of their family, one retains gratitude, affection, and in the case of the *goy*-child, some of the neuroses. All the more neurotic and poignant when Jewish parent-friends never admit to the Jewishness which still oozes from every pore despite their conversion to this or that incongruous branch of religious faith. It is a situation spread with the thinnest ice.

Isolated at Castle Hill we played the radio indiscriminately: Vera Lynn at breakfast encouraging us to eat 'Creamoda', the young Joan Sutherland competing in Mobil Quest, Carl Orff, Yma Sumac, Marjorie Lawrence in her concert versions of *Electra* and *Salome*. We had seats for the Lawrence performances, but as often happened towards the night, I was bedridden with bronchial asthma.

Whenever I had to take to my bed a little girl from the family whose bull we patronised would come down to milk. There were always the cows to worry about, goats later on, as I lay in bed wheezing and hawking, scribbling more often than not, breaking off to change the gramophone record.

Bronchitis, Menuhin playing Bartok's Violin Concerto, and a virulent review of *The Tree of Man*, helped me resolve the death of Voss. I had not felt up to it before. Suddenly I was injected with adrenalin enough to hack off the head.

The brassy glare of summer seemed to make the furniture in that small house crowd together and glare back. Carpets grew dustier, more noticeably Wilton and Axminster. Winter on the other hand emptied the interior: we were living in a freezing, grey tomb. After a few years we had stoves installed in the fireplaces, not so much for warmth (we liked to think ourselves austere, and certainly we were hard up) but the stoves stopped the possums coming down the chimneys late at night when the fire burnt low in the grate. There always followed a steeplechase as schnauzers pursued a possum, and next morning sooty claw-marks on walls and paintwork, and urine patches on the Genoa velvet suite.

In blocking the flues, the stoves created a fine winter fug, along with Bruckner, cheap sherry, and smells from the grill cooling in the kitchen.

The kitchen stove, antique electric, had Queen Anne legs like so much of the furniture in the house. We had many cooking accidents before learning how: there was the day Rosemary Dobson Bolton brought her first baby to lunch and the oven in which I was grilling the baby's chop caught fire; the Christmas dinner during a heatwave when the pair of drakes (or swans) we had bought from Mrs Poulter bounced on the lino before I dished them up; there was the whole *coq au vin* I spilled on the floor, but mopped up, schnauzer hair and all, and served John Gielgud. Although I say it, that *coq au vin* was about the best I have tasted.

Food was crude and rude, like everything else in the Castle Hill days, when we were still making our mistakes.

The novels? Possibly. I never make the mistake of re-reading a novel once it has gone to the printer. That is why I can't answer people who ask me the meaning of line 7 page 18. Playing safe? Perhaps. On the other hand I have to get on with what has still to be written.

The mirror in the bathroom at 'Dogwoods' had a flaw in it like a faint birthmark. I associate the bathroom with Manoly's voyage to Greece in 1958. He was away what seemed like years, but the break had to be made to exorcise the homesickness, the doubts, the longings most migrants suffer from. I had the bathroom renovated. A sulky youth laid approximately six tiles a morning, then sat around listening to a transistor. I couldn't wait for him to knock off and leave me to a different kind of solitude.

While on my own I was growing vague. Once in the middle of the night I found myself about to piss in the bathroom waste-paper basket. I grabbed at food, I guzzled drink. The dogs were my only comfort, Solomon sleeping at my feet, Lottie snuggled into one armpit.

Some critics complain that my characters are always farting. Well, we do, don't we? *fart*. Nuns fart according to tradition and *pâtisserie*. I have actually heard one.

Until painting this self-portrait I had never tried to draw the graph of my religious faith. There were the Gentle Jesus prayers gabbled from behind Gothic hands to Mummy or Nursie in the days when we were Pretty Baa-Lambs. My pagan moment chanting a home-made hymn in the latticed summerhouse in the lower garden at 'Lulworth' which so shocked my worthy areligious dad. The services at St James's in the city under the spell of Mr Micklem's ascetic icon and the fainting ladies carried out by vergers and propped on chairs in the porch, at the same period Una de Burgh (the Bad Fairy) who wanted my voice for the choir. Confirmation at Cheltenham and waiting for the miracle which didn't happen.

I threw it all off in my late teens. Then, and in my early manhood, I was too egotistical, too sensual, to consider spiritual matters. As an Australian, perhaps too materialistic – though God knows, many of the more enthusiastic Australian evangelists, Methodist and C. of E., are materialistic enough.

Greek Orthodoxy's straitjacket prevented Manoly ever losing his religious faith. During the brash barren years of what I saw as intellectual and sexual freedom he was my ethical if not my spiritual guide. Because I loved him and was grateful to him, perhaps in the end he influenced me spiritually too. Looking back it is difficult to tell. I can only recall my disillusionment and despair for the wrong turning I felt my life had taken when I came back to Australia.

My work as a writer has always been what I understood as an offering in the absence of other gifts. *The Aunt's Story*, my first published work after settling at Castle Hill, was considered freakish, unintelligible – a nothing. You only had to pick up a library copy to see where the honest Australian reader had given it up as a bad job. I brooded after that. I considered giving up writing altogether, before starting on what was in some ways the even more calamitous *Tree of Man*. Living amongst our weeds, in what was an asthmatic hell, dragging the milk out of a couple of cows and selling their thick, illicit cream, I believed in nothing but Manoly, whom I had brought on a wild goosechase to the other side of the world. When our funds were at their lowest, he went out mowing lawns and stripping schnauzers. Members of 'old families' and a lady whose husband had made money out of saucepans treated him with condescension. On the other hand a widow fell in love, and bicycled down regularly for cream and conversation.

The seasons we experienced ran through every cliché in the Australian climatic calendar: drought, fire, gales, floods along the road at Windsor and Richmond. During what seemed like months of rain I was carrying a trayload of food to a wormy litter of pups down at the kennels when I slipped and fell on my back, dog dishes shooting in all directions. I lay where I had fallen, half-blinded by rain, under a pale sky, cursing through watery lips a God in whom I did not believe. I began laughing finally, at my own helplessness and hopelessness, in the mud and the stench from my filthy old oilskin.

It was the turning point. My disbelief appeared as farcical as my fall. At that moment I was truly humbled.

We both began an exercise in organised humility. There is nothing remoter from Greek Orthodoxy than Sydney Evangelical C. of E., but in its hatred of Rome the Eastern Church had accepted Protestant overtures. A Greek living at a distance from a church of his own faith might attend the local C. of E., as Manoly did during our years of trial and error.

Every Sunday we set out for early communion as it did not interfere with our activities about the 'farm'. Built in the early days, the church at Castle Hill had accumulated the kind of Victorian and Edwardian bric-à-brac with which prosperous Australians express their gratitude for God's recognition. Kneeling in this church, under a succession of worthy and not so worthy rectors, in winter frost or the cool before a summer blaze, perhaps I was awaiting unconsciously one of the miracles which had not occurred after confirmation. Secure in a more elaborate tradition Manoly was less expectant or more sceptical. Like any expatriate he was not responsible for the farce he had dropped into. I could not protect myself as he did from the bigotry we found. We withdrew after the rector of the day declared it sinful to guess the number of beans in a jar at the annual church fête.

For a brief space we tried driving to Sydney to the service at Christchurch St Lawrence, unintelligible to both of us, though there were some nice moments of theatre as the acolytes, including a young Chinese, strolled among the faithful weaving veils of incense. Surprisingly, it was the esoteric element which caused our withdrawal – and the presence of the Bad Fairy from my childhood at St James's, now in charge of vestments at Christchurch. So each of us retreated into his private faith, and there we have remained. Each respects what the other believes, though Manoly, I think, disapproves of my erratic spirit, chafing free, rejecting tradition. Is it ever possible to believe entirely in somebody one knows by heart, who is, at the same time, the one it is impossible to know?

My inklings of God's presence are interwoven with my love of the one human being who never fails me. This is why I fall short in my love of human beings in general. There are too many travesties of an ideal I am still foolish enough to expect after a lifetime's experience, and knowledge of myself.

In my own opinion my three best novels are *The Solid Mandala*, *The Aunt's Story*, and *The Twyborn Affair*. All three say something more than what is sacred to Aust. Lit. For this reason some of them were ignored in the beginning, some reviled and dismissed as pornography. After years two of them were accepted; it remains to be seen what will become of *The Twyborn Affair*.

Strange to think *The Solid Mandala* was ever considered pornographic, yet an Australian professor told a friend it was the most pornographic novel he? she? had ever read. One wonders where he or she spent his or her literary life before *The Solid Mandala* appeared.

It was begun during the last of our eighteen years at Castle Hill. The painter Lawrence Daws had given me Jung's *Psychology and Alchemy*, which had a great influence on me. It projected me into my *Solid Mandala*. Jung's teaching also bolstered me up during a wavering of faith on realising I could not accept the sterility, the vulgarity, in many cases the bigotry of the Christian churches in Australia. Manoly seemed secure inside the structure of Eastern Orthodoxy. I had nothing from my upbringing in a kind of social C. of E. (a visiting card on the pew, clothes outgrown or no longer fashionable sent off to the jumble sale, a grateful rector and his wife calling to express gratitude for patronage.) So I evolved what I think Manoly has always seen as my non-religious or mystic circus.

Some of all this rubbed off on *The Solid Mandala*. It is a book full of ambivalence and unease – transitoriness. The trees we had planted, tended, and loved, were encroaching on the house, making it darker. We knew that we must leave 'Dogwoods'; it was impossible to continue living in what had become a suburb, when our interests – music, theatre, film, friends – were concentrated in the city. But to make the break was hard. In my case, I hated what I loved. In Manoly's he loved more deeply, because he has less hate in him, the trees which would inevitably be chopped down.

I was also afraid to sever the spiritual roots I had put down in that originally uncongenial soil. Would it be an end to my writing? Bad enough when we uprooted ourselves and flew to Europe, for me the first time since leaving it fifteen years before. We must have looked a comic pair, slung with hand luggage, wearing unfashionable trousers and hats, as we staggered out along the tarmac, waving back to the friends who were seeing us off. How much worse the other, permanent uprooting seemed, as I sat in the dark dining-room at the centre of the house, writing the novel which could be my swan-song. It is not surprising *The Solid Mandala* was infused with an amount of fatality and foreboding.

I see the Brown brothers as my two halves. Arthur might have been a portrait of my cousin Philip Garland if Philip's childish wisdom had matured; instead he was admitted to an asylum in his teens, and remains in one to this day. Waldo is myself at my coldest and worst. The Browns' neighbour Mrs Poulter grew out of our actual neighbour Mrs H., though the latter was a two-dimensional character beside Arthur's earth-mother and goddess. Dulcie Feinstein has both the goodness and the smugness of a fulfilled Jewish acquaintance.

When the time came to make the break with Castle Hill, my cousin Eleanor Arrighi was lunching with us one day. From being Nelly Cox of Mudgee, Eleanor had become a showgirl, later the wife of an Italian diplomat. At the period of which I am speaking, she was widowed, settled in Sydney, and making a living as an estate agent. At lunch I took the plunge and asked her to find us a house in Sydney, 'it's got to be near a park for the dogs.' She said, 'I'll take a drive along Martin Road, Centennial Park, on the way home.' In doing so she found a FOR SALE sign at Number 20.

That is how we came to live at 20 Martin Road, in a house built in 1912, the year Manoly and I were born.

The great upheaval took place on October 13th 1963. We discarded two-thirds of what we possessed – not nearly enough. The rest was brought from Castle Hill, furniture and paintings in removal vans, unhappy dogs, terrified cats, and passive plants carried by David Moore* and Manoly in their station-wagons. John Young* helped us unpack. Klari Daniel* brought us lunch.

I had moments to myself in this unfamiliar, still forbidding, freshly renovated house. When helpers had left for another load, and removalists had not yet arrived, I sat at the broad window-sill in the dining-room and took up *The Solid Mandala* where I had been forced to interrupt it, wondering whether I should be able to join the threads where they had been broken. My first attempts at doing so in the deserted house were pure, if fearful, bliss.

So we began our life in Martin Road, amongst the smells of fresh sawdust and paint, and the dust of years we had brought with our belongings from 'Dogwoods'.

I am sure our ghosts will always haunt Showground Road, the dark little house with cracks in its walls and white ant in its foundations. Those who are psychic or unhappy may still catch a glimpse of us running out naked by moonlight amongst the regimented boxes which now stand where the trees were cut down. Perhaps my laughter will be heard on Nobel Avenue (true!) where I fell on my back in the mud beside the cow-bail, cursing a God in whom, I realised, I must believe after all.

Again in Martin Road nobody wanted to know us at first, but it did not matter; nobody knows anybody, whether in the beginning or the end — which is one of several reasons why I am writing this book. As always, the neighbours must have found us odd — a couple of men living together. A writer, moreover, might write about you.

The dogs adapted themselves very quickly, a dog's home being where the masters are. The pugs Fanny and Ethel accepted the situation with the usual grave pug wisdom, whimpering only fitfully. More delicately sceptical, the minpin Lucy seemed to turn up her nose at everything she smelled in city grass. For the cats the move was real severance from their life-roots. They spent a fortnight in the garage before venturing out through the hatch to rout invaders and stake their territorial claims — Pearl the cold white narcissist in front, the back yard shared by black, hissing, increasingly crazy Dilly, and Dixon, a white picked up some years before amongst the Chinese garbage of Dixon Street. Dilly's niece, Cobby, the diplomat among cats, was accepted by the others wherever she chose to show herself. A blue who started life as Cobweb, she lived with us for twenty years till cancer of the ear forced us to put her down. We had brought her up on an eye-dropper, tickling her stern with cotton wool to encourage her excretory functions after her mother had been squashed flat on Showground Road. Perhaps as a result of the personal touch in her upbringing, Cobby became the most devoted cat I have known. In her old age she adopted Ugly, a tortoiseshell kitten who moved in to escape torture by neighbouring children. Cobby was as devoted to the not surprisingly neurotic Ugly as she was to their two neurotic owners. Throughout her life she remained the spirit of calm.

Built in the year of our birth, the house in Martin Road might have been put there for us, and Nelly Arrighi led there by fate. The inevitable black moments we have lived through, and would have lived through anywhere, have been offset by an atmosphere of compatability contributed by the house. Though it grows larger every year, and more unmanageable for two old men, I hope I shall die in it, at the risk of the event's taking place in dust and squalor.

The garden was non-existent when we came: three gumtrees in front and a sand-patch behind. The aboriginal owners of the dunes which became a middle-class suburban precinct bounded by Moore and Centennial Parks, called the summit Mount Meroo. (That our house stands on this rather insignificant hillock flatters the egotist in me since I read that the Mount Meroo in Central India is considered the centre of the world.) It took Manoly seven years to establish a garden, part native, part European, on our dune. He is still adding strokes of colour and removing misguided formal details. It could go on for ever. There is no end to a garden, unless the bulldozer; just as a writer's hand is halted only by death or paralysis – or worse still, the rambling side-tracks of senility.

When we had settled down to living at Centennial Park I prepared to write the two novels which belong to Sydney. *The Vivisector* was already forming at 'Dogwoods', but in spite of time spent in my native city during childhood and youth, and frequent visits while we were at Castle Hill, I had to feel Sydney round me, day and night, in my maturity, before I could undertake the novel. *The Eye of the Storm* came to me crossing Kensington High Street, London, after a visit to my mother at her flat in Marloes Road where she was lying bedridden, senile, almost blind, tended by a swarm of nurses and servants. I knew I would write this novel about some such old woman at the end of her life, but in a house in Sydney, because Sydney is what I have in my blood. No more than this flash of prescience at the moment of crossing Kensington High Street. An incident changed and elaborated in the novel was the attempt Sue and I made to persuade our mother to move in with the Blue Nuns of

Holland Park. We had investigated this house with its large dignified rooms and peaceful atmosphere. It appealed to me so much I felt I could have moved in there and then and detached myself for ever from the world. Ruth refused to contemplate the idea, and died in her flat after I had returned to Australia. I have always had the guilty feeling that our suggestion may have killed her, not so much the prospect of leaving possessions she could no longer see, as the thought of dying surrounded by Roman Catholic nuns. Elizabeth Hunter's children in the book are more vindictive in their efforts to put her in a home. There was never anything vindictive in Sue, nothing of the Princesse de Lascabanes, though plenty of the latter in me, and as a frustrated actor, I can recognise myself in the vulgar egotist Sir Basil. I have sometimes wondered whether Dorothy and Basil were avengers I summoned up from the depths. Was my suggestion that Ruth should move in with the Blue Nuns my unconscious retaliation for the years of purgatory to which she condemned me at Cheltenham? I hope not, but it could be so.

Back to Sir Basil: Joe Losey once told me Dirk Bogarde considered Basil would never have been knighted because he was such a vulgar man. Looking around me, then and now, particularly in Australia, this opinion has given me many a laugh.

The Vivisector, which came first, is about a painter, the one I was not destined to become — another of my frustrations. I had imagined that if I could acquire the technique I might give visual expression to what I have inside me, and that the physical act of painting would exhilarate me far more than grinding away at grey, bronchial prose. This could be the delusion of a writer who has always resented having to write. Some painters have told me that Hurtle Duffield is not a painter, others that he is. Throughout my writing life I have encountered fiercely contradictory judgments: that Himmelfarb is/is not a Jew; that I know everything/nothing about women; that what I write illuminates, or on the other hand, that my novels are incompre-

16   Suzanne with her first child, Gillian, England 1940.

17   P.W. at 'Dogwoods' with Tom Jones, sire of many whites, and
Solomon the perfect companion.
18   P.W. with illicit cow, 'Dogwoods', Castle Hill.

19   'Dogwoods', Castle Hill, front view.

20   'Dogwoods' from back.

hensible, boring rubbish. But I expect any writer who takes risks has had this battle fought out over his body, live or dead.

Whether Hurtle Duffield is or is not a painter, I see him as a composite of several I have known, welded together by the one I have in me but never became. Setting out to portray a convincing artist, I wanted at the same time to paint a portrait of my city: wet, boiling, superficial, brash, beautiful, ugly Sydney, developing during my lifetime from a sunlit village into this present-day parvenu bastard, compound of San Francisco and Chicago. I had a lot of exploring to do. It was not so much research as re-living the windswept, gritty, or steamy moods of the streets, coaxing dead-ends, narrow lanes, and choked thoroughfares to release those voices, images, emotions of the past, which for my deplorably atypical Australian nature evoke guilt rather than pleasure.

The puritan in me has always wrestled with the sensualist. As a child I felt ashamed of my parents' affluence. I was aware of a formless misery as well as material distress the other side of the palisade protecting the lives of the favoured few. For that reason I have never been able to enjoy what any 'normal' member of my parents' class considers his right. What is seen as success, my own included, has often filled me with disgust. No doubt the 'normal' members of the affluent class will pounce on this confession as explanation of what they think a distorted viewpoint in all I have written; while to me the refractions from that many-sided crystal, truth, are more diverse than they would have been had I remained blinkered by the values of the Australian rich.

After settling at Centennial Park we began, you might say, going into the world. I had to find out how the rituals of a fashionable society had changed since my parents' day. Not all that much, it seemed at first. There were still elderly maids in starched caps serving at table in ladies' clubs. One old girl, May, remembered working for my mother at 'Lulworth'. It was a departure from convention when some of the relics went out cooking and serving meals in private houses. The menu was always the same: vichyssoise, beef fillet, and exquisitely sliced French beans. They were touching, these old women, their arthritic hands offering platters, jogging your elbow if you hadn't responded to their offerings. There was one I remember called Myrtle Park, a frail, Dickensian spinster I always enjoyed meeting again over my shoulder.

There were grander occasions in the grander houses of those who collected celebrities. On one I sat next to Stravinsky at dinner. We got off to a flying start after each had confessed he was an alcoholic. I can't remember much of what we talked about: snatches of gossip, some of his Diaghilev ballerinas I had seen dancing for de Basil in London in my youth. I have always found it difficult to talk to musicians; a musician is his music. But I think Stravinsky and I enjoyed sitting side by side downing the drink we had in common. After all, I had listened to his music, and he claimed to have read some of my novels. There was little more we could expect of each other. Along the table Robert Craft kept an eye on us, from time to time trumpeting about something or other in case we had forgotten he was there, and farther down there was Madame Stravinsky, like a Russian version of an old English sheepdog. It must really have been an evening of shocking boredom for the Stravinsky Trio. A few nights later I went to a concert at which Craft conducted an American collegiate version of a Stravinsky work, before the little showman himself, shrunken, wrinkled, supported by a stick, hobbled out to marshal his *Apollo Musagettes*. It was an august occasion. Whether it quite warranted the applause, those present clapped themselves silly; musical audiences always suffer from the clap.

As we raced through the 'Sixties into the 'Seventies the social climate changed: ladies of a higher social level began cooking for their equals, their inferiors too, if the money was there. Money became everything, vulgarity chic, the crooks got off provided they were rich enough. Knighthoods could be bought more easily than ever as inflation rose and the British Monarchy commuted regularly with an eye on Australia as their remaining bolthole.

During the 'Seventies I withdrew from circulation. I had got to know the habits of contemporary rank and riches. Some of those belonging to our affluent society would argue that they had dropped me for being a traitor to my class. I made no secret of my political allegiance. I spoke out in public. In any case I have always seen the artist as classless. If what I have said about the Australian so-called Better Class sounds priggish and hypo-critical, I should add that my own family belonged in the cate-gory of new-rich when they came to Australia from Somerset as yeomen-farmers generations earlier and were granted great tracts of land which they proceeded to farm, professionally and profitably. As a result of their success they began building Edwardian mansions to replace their simple, early homesteads. Their imported motor-cars were the equivalent of today's Mercedes, Jaguars, Porsches, and Ferraris. Austere in many ways, my forebears were also flash in what has become established as the Australian new-rich tradition. The distinction lies in the fact that my father and his brothers were honourable men who would not be divorced from their principles. My dowdy aunts had a moral core which could not be faulted. Even my more pretentious, more elegant mother would never have shed her principles. We were brought up never to blow our trumpets, talk about money, live beyond our means, but to give quietly. (Uncle Arthur, the meanest of the Whites, gave on the quiet, I have since discovered.)

So, for all I know, the creatures squirming in today's social pool, ladies with their lip pencils and exposed nipples, shady knights and captains of self-advertisement may indulge in quiet giving between jetting to and fro, their variations on adultery, and appearances in the courts of law. It is possible. Double values abound amongst those I used to respect; and as for myself, I have never disguised a belief that, as an artist, my face is many-faceted, my body protean, according to time, climate, and the demands of fiction.

All the houses I have lived in have been renovated and re-furnished to accommodate fictions. The original structure is there for anybody who knows: 'Lulworth' for *Voss*; 'Dogwoods' for *The Tree of Man* and *The Solid Mandala*; Martin Road for *The Eye of the Storm*; the cottage, the homestead, the sheds, the dunny at 'Bolaro' for *The Twyborn Affair*. In some cases it has not been so much architecture as atmosphere which has transferred the house to the page. The spirit of the stairs at Martin Road as I groped my way down most mornings in the dark conveyed the interior of a Paddington house I saw as Duffield's without ever having been inside. In the theatre of my imagination I should say there are three or four basic sets, all of them linked to the actual past, which can be dismantled and re-constructed to accommodate the illusion of reality life boils down to.

I sometimes wonder how I would have turned out had I been born a so-called normal heterosexual male. If an artist, probably a pompous one, preening myself in the psychic mirror for being a success, as did the intolerable Goethe, inferior to his self-abnegating disciple Eckermann. My unequivocal male genes would have allowed me to exploit sexuality to the full. As a father I would have been intolerant of my children, who would have hated and despised me, seeing through the great man I wasn't. I would have accepted titles, orders, and expected a state funeral in accordance with a deep-seated hypocrisy I had refused to let myself recognise.

As a woman, I might have been an earth-mother, churning out the children I wanted of my husband, passionate, jealous, resentful of the cause and the result, always swallowing the bile of some insoluble frustration. Or I might have chosen a whore's life for its greater range in role-playing, greater than that offered an actress, deluding my male audience of one into thinking I was at his service, then flinging back at him the shreds of his self-importance as he buttoned up. Or else a nun, of milky complexion and sliced-bread smile, dedicated to her quasi-spiritual marriage with the most demanding spouse of all.

Instead, ambivalence has given me insights into human nature, denied, I believe, to those who are unequivocally male or female — and Professor Leonie Kramer.* I would not trade my halfway house, frail though it be, for any of the entrenchments of those who like to think themselves unequivocal.

In fact sexuality refreshes and strengthens through its ambivalence, if unconsciously — even in Australia — and defines a nation's temperament. As I see it, the little that is subtle in the Australian character comes from the masculine principle in its women, the feminine in its men. Hence the reason Australian women generally appear stronger than their men. Alas, the feminine element in the men is not strong enough to make them more interesting.

One English critic finds it a serious flaw in my novels that my women are stronger than my men. I see nothing anomalous in this imbalance; it arises from a lifetime of observing my fellow Australians, in closest detail my own parents when I was young.

Where I have gone wrong in life is in believing that total sincerity is compatible with human intercourse. Manoly, I think, believes sincerity must yield to circumstance without necessarily becoming tainted with cynicism. His sense of reality is governed by a pureness of heart which I lack. My pursuit of that razor-blade truth has made me a slasher. Not that I don't love and venerate in several senses — before all, pureness of heart and trustfulness.

Journeys

To the Holy Mountain and the Past

We started off apprehensively from the obsolescent Athens railway station. The train which would meander on through Europe already smelled of stale air, cigarette butts, excrement and urine. Obsolescence and fatality loom around each setting forth in Greece. Manoly is unhappy because he feels responsible for it, as I am in Australia for all I want to see improved. We have spent our lives apologising to each other, often mutely, as we catch sight of the cockroach skittering in the shadow of the skirting-board.

There are many such moments in Salonika. Ancient ruins and Byzantine mosaics fail to distract the mind from the air of Slav menace on the city's northern rim. Our hotel would have accommodated the more sinister sequences of some cloak-and-dagger 'B' film. During the night, a French letter in the lavatory bowl refused to be flushed by either of us. Then in the morning we had our first glimpse of Olympus through the haze above the curving bay. Any true Grecophile will understand when I say that the unsinkable condom and the smell of shit which precede the moment of illumination make it more rewarding when it happens.

We set out for the Holy Mountain in one of the cranky Greek buses, checking and re-checking the papers we had extracted from a (holy) department in Athens. Our hand baggage was lumped round our feet, the larger pieces corded to the roof under a tarpaulin. It was not till scaling cliffs to reach monastery gates which close at sunset that we realised we should have come with knapsacks, a change of underpants and socks, and a couple of Penguins. Anyway, here we were, bumping in the rickety bus towards our port of embarkation, a dispiriting village where plastic and the juke box had invaded the *kapheneion*. We settled down to wait.

The caique the pilgrims eventually boarded (you do in fact become a pilgrim) chugged along parallel to this sandy finger of monotonous land. None of the splendours associated with the Holy Mountain reached out to grapple the expectant soul. So we settle down again. (Settling down is a condition peculiar to journeys through Greece; hopes are damped as arrival is postponed; it is a country for masochists like myself.)

At Kariyes, capital of the Church State, we were billeted in concrete dormitories in the equivalent of a hostel. Squalor abounding. Next morning the Church bureaucrats fossick through our papers adding a few more stamps. We stroll around and inspect the souvenir shops at the wrong end of our visit. We buy three little wooden icons like squares of gingerbread carved with saints. Outside an almost deserted Russian monastery, an inflated Fabergé object in chipped viridian, a monk with a look of Verlaine expressed hostility by screaming at us. Time and again we found that Orthodox monks scream like enraged queens. There was nothing we could do to calm this aged Russian, so we walked away, and soon it was time to bring our shamefully superfluous luggage to the mules we had hired to carry it down to Iviron and the caique which would take us on the next stage of our journey — to the Grand Lavra.

Much as the bus from Salonika bumps you over grey, potholed roads, the caique tosses you on blue waters past cliffs with monasteries attached like the nestings of ingenious birds or insects. What could only have been projecting dunnies constructed by monkish spiders added to the grotesque beauty of the Holy Mountain's architecture. The exception is Vatopedhi, a Byzantine version of a Harrogate hotel. (The only time I saw Harrogate I was with my family at the same hotel and historic moment when Agatha Christie checked in without her memory.) During our stay at the Lavra, we returned by mule and spent a day at Vatopedhi. The florid buildings clustered round the main court attempt to exclude the world, but even serenity is of a worldly order at Vatopedhi. In spite of relics and the presence of monks, that first impression of Harrogate was replaced by one of Brighton. We were given lunch on our own in a private dining-room, a meal in keeping with its comparatively worldly surroundings, and certainly the only civilised food we tasted in the time we spent on Mount Athos. We would not have minded shedding our fellow pilgrims and sinking into the feather beds and hot baths Vatopedhi must surely have to offer.

As pilgrims we were an incongruous lot: Manoly and I with our excess luggage; the trumpet-player from an Athens night-club whose mission was to bring back a leaf from the miraculous vine at the Romanian monastery of Hilandari to help his childless wife conceive; a young policeman with progressively smellier socks; a Jehovah's Witness with a badge in his beret who provoked a quarrel almost every time he opened his mouth; a restaurant proprietor from Thebes who expectorated continually, and drove his phlegm into the Grand Lavra tiles. What we had in common was our pilgrimage, our exasperation at the slowness of it, and anxiety that we might not arrive before the monastery closed its gates at sunset. When we did tie up in the little harbour below the Lavra the sun was already setting. There began a most infernal scramble up a precipice, myself wheezing, the Theban coughing and hawking, all of us by now stinking, ourselves almost ruptured by our suitcases, the typewriter I never once laid finger on throughout the journey tinkling feebly as it was mishandled. (The reason for our ridiculous overloading was that in our simplemindedness – or snobbery – we had brought along clothes we thought might be appropriate to Constantinople and Smyrna, something in which to pay tribute to Manoly's Byzantine ancestry.)

To the disappointment of the gatekeeper-monk, we just made it at the Lavra as the sun went down.

In travel anywhere I have discovered you arrive almost always at the wrong moment: too hot, too cold, the opera, theatre, museum, is closed for the day, the season, or indefinitely for repairs, or else there is a strike, or an epidemic, or tanks are taking part in a political coup. At the Lavra we coincided with preparations for celebrating the millennium. Nerves were jittery, tempers ran high as imported workmen attached lath, plaster, and masonite to stone walls which had withstood the ages. Never wanted, but accepted in bad grace, pilgrims were more than ever superfluous to the lives of the monks. One speculated how the Church State would react to the King of the Hellenes, who was expected as its civil guest of honour.

Our first evening started amiably enough with the monk in charge of pilgrims offering thimble-sized glasses of ouzo and squares of *loukoumi** on a balcony outside our quarters. The situation deteriorated as the evening wore on, through a meal of slimy *kritharakia** and naked haricot beans (hadn't we chosen the ascetic way?), the dirty wash-house with broken panes, the expectably filthy lavatories, arguments in the dorm before lights-out, and the policeman-pilgrim's overpowering socks.

On the Holy Mountain, even more than in Greece itself, waiting is a condition you have to get used to. Impatience always let me down. Manoly who is sweet-natured, Greek Orthodox, and a fatalist, came out of it with greater dignity. But whatever our virtues or our flaws, we continued waiting till the monk in charge consented to show us a vestment, a missal, or the relic of his saint.

It would have been more humiliating if the worldliness and bad tempers of some of those holy monks had not matched my own. As we made our way humbly through the stone labyrinth, barely sustained by our ascetic diet of *kritharakia* and naked beans, delicious smells of cooking drifted out of private suites, glimpses of decor offered themselves between the opening and closing of a door, exquisite acolytes tittuped on some errand for an imperious superior. I was reminded of the monasteries of Athos when told of a slogan outside an evangelical church on the South Coast of New South Wales, 'Churches are hospitals for sinners, not hotels for saints.'

Hotels for the 'saints' they house are what some of the less ascetic monasteries of Athos have become, and no doubt the sumptuous Vatopedhi, that Harrogate Hydromajestic-cum-Brighton Pavilion, will end up as an actual hotel for rich, homogeneous tourists of both sexes rather than the assortment of male eccentrics we found ourselves to be during our brief pilgrimage.

And yet, even the most exasperated and temporarily scornful pilgrim can be raised from the slough of his cynicism by the silence of a pine forest, or when, under the blue and gold crag of the mountain itself, an icon in its details of jagged rock, tufts of grass, the florets of a stubborn plant rendered with Byzantine formality, an anchorite of true vocation, looking like a rusty old black umbrella, stumbles down a stony track ahead of his donkey – all contribute to blot out the impression of contaminating luxury in the cells of the hierarchy.

We, the novice pilgrims, are roused from our restless sleep under hairy blankets in dream-infested dormitories by monks whirling wooden klaxons or *semantra* in the alley below to summon us to early devotions. We struggle up and into our clothes. The church still smells of cold wax and stale incense, though live candles are beginning to warm it through. A liturgy is grinding into gear. We look bleary and confused, huddled together in upright self-effacing groups, till giving in to the voluptuousness which lies the other side of asceticism, we dare loll against the wooden supports the walls of an Orthodox church offer those who are spiritually insufficient. As voluptuaries, lulled and lolling, drunk on the fumes of fresh incense and the flow of immemorial words, our souls respond, or so it appears. If I cannot truly claim that the other drunken, bleary, unwashed, unshaven faces share my experience, it is immaterial as my spirit ascends through the branches of the liturgy and loses itself at last in the cloud of bluish smoke.

After several hours of worship, hunger and boredom begin to gnaw at ecstasy. Later that day we bought a tin of Spam at the gateway shop and surrendered ourselves to the sin of gluttony. It was a delicious relapse, equal to those moments in the

Western Desert in wartime when you broached a tin of bully or Meat and Veg. Lacerated fingers were not the least part of it.

Lacerations alternating with visions: is this what hooks the more perverse Grecophile? Whether lolling against worm-eaten woodwork in ancient Byzantine churches, lulled by the liturgy he is hearing, or seated on a rush-bottomed chair, seduced by the feminine slither of the sea, its lip extending till almost sucking at his toes, while a golden hen scuffs round the iron roots of the table, and the beads of the *komboloyi** restlessly cannoning off one another draw some of the nervous fever from the hands at play, the Greek is never wholly unconscious of the echoes from a torture chamber in which his psyche is a permanent victim. Initiated into cruelty by Turk and German he is not above torturing his fellow Greek, which rebounds on him as self-torture. Anglo-Saxons who, in general, have not experienced the cruelty of foreign oppressors are more inclined to indulge in games of mental cruelty, while Australians in particular torture their fellows by apathy.

During our stay at the Lavra a storm arose which lasted several days, so that we gave up hope of visiting the monasteries on the other side of the promontory. A caique could not have rounded the cape even if one had arrived in port. We felt both frustrated and relieved, for each of us by now had memories and insights enough of the ascetic life on the Holy Mountain. To be perfectly honest we could not wait to get away. Each afternoon we walked down to the harbour hoping to see our rescue caique riding the storm-tossed waters, but were disappointed. As were the monks. One evening while engaged in the desperate watch for a boat, I just missed having the door slammed on me at sunset by a big black bearded monk-queen.

In the end it got to be like boarding school all over again, ticking off the days on the calendar. But the caique did arrive at last, and we were carried back to what is variously labelled, as 'life', 'the world', 'freedom'. Our visit to the Holy Mountain was an experience which brought out the worst and the best in me. I felt physically battered, spiritually leaner, which was all to the good. Manoly, who is more best than worst, was possibly less changed, but I cannot truthfully say. Most important, we could continue on our journey to the Polis – Constantine's City. And the trumpet-player had bagged the miraculous vine-leaf which would help his wife conceive.

So we chugged away through the glare and wind, our faces textured in some cases like cured bacon, in others the skin of half-fried sausage. At least Jehovah had been silenced, either by a state of grace or the hostility of his fellow pilgrims.

One last evocation of sanctity and faith as we travelled towards Salonika, churned around in our atrocious bus by deep ruts in the tortuous road, and occasionally shot against the roof. A policeman on leave from a village en route discovered he had lost his wallet, in it not only his money, but the papers which established his identity and ensured his self-respect. Every passenger in the bus was to some degree infected with the young man's hysteria. We searched as though each of us was respon-sible for the lost wallet, and reinstatement of one's ethical, not to say spiritual self, depended on its recovery. Some way down the aisle a small shy Romanian monk from the Monastery of the Miraculous Vine murmured that he was praying to St Phanourios for recovery of the lost wallet. No one discouraged his prayers, but nobody seemed particularly encouraged. A little farther on, the policeman stopped the bus in a village to telephone his station for advice. None seemed forthcoming. We trundled on, some of us still twisting and turning from time to time to do our duty by the lost wallet. Its owner in his frenzy smoking Greek cigarette after Greek cigarette, was visibly saturated at the armpits. The monk, his eyelids meekly lowered, remained withdrawn presumably inside the walls of prayer. Several kilometres on, the wallet's owner shouted he had found it wedged between the chromium base of his seat and the shell of the vibrating bus. Such joy! All hail St Phanourios! Gold teeth flashed and glistened, the Panayia jiggled appreciatively against the windscreen, muscular and arthritic hands clapped the Romanian on the back for bringing off a minor miracle.

The rest of our journey was too much anti-climax. When we dispersed at the terminus we were naturally preoccupied with practical moves in the immediate future. It was the *Boule-de-Suif* situation in a contemporary Greek setting. Vaguely one noticed the monk retreating down the long perspective of a Salonika street carrying his little case, vaguely one hoped some extra-monastic pleasures were awaiting him at his destination.

Towards the Polis

I can't remember for what reason Drama became one of our staging posts; perhaps because the name on the map had always intrigued me, or because it was a convenient point from which to join the Stamboul Express. It turned out to be one of those provincial towns not so very different one from the other, from Tripolis to Kastoria, from Patras to Chalkis: the same ennui, the promenade at dusk, the same melancholy restaurants for resident bachelors and itinerants, with food afloat in baths of tepid oil. Except that in the north you are more conscious of a Slav cloud drifting down from the frontier. In addition to the Slav cloud over Drama, there are the gipsy faces. It is a town of coppersmiths, and the streets in which their booths are concentrated have something of a gipsy fair. We would have liked to buy up a truckload of Drama copperwork, and for ever after regretted having to clean the stuff.

If certain streets were copper-coloured, Drama's predominant colour was grey. Pools of steely water in the centre of the town were renewed by springs plopping and bubbling under them, disturbed also by shadowy trout, some of which appeared on our plates during a meal. In my more orthodox Christian phases I was always repelled by the idea of transubstantiation. In the same way I have often been repelled by eating the flesh of my favourite fish, one of the most exquisite of living creatures, and a work of art when lying dead. We tried to enjoy Drama, but the overall grey and its damp Slav aura quenched our hopeful enthusiasm.

Because the train passed through late in the day, we had taken a room in which to keep the siesta. Except that the hotel at Drama smelled slightly mustier than the others we had experienced throughout provincial Greece, it was much the same: a mild stereotype art deco in the foyer and on the stairs, modernistic furniture in bedrooms which suggest they have been closed for weeks. Open a drawer and the scent of imprisoned face-powder escapes to assault allergic nostrils. The sheets on which we barely rested were, not surprisingly, damp and grey.

While everything was more or less as we expected, we were suddenly roused from our restless dozing by a Siamese cat bounding in from the balcony adjoining ours – the first human touch in Drama. We had to investigate, and found the cat belonged to an Athens cabaret artiste doing the provincial *boîtes*. The young woman's respectability appeared as inviolable as the helmet of her blonded hair, but no doubt some black, hairy Greek was waiting to ruffle pubics which did not match the helmet, and pay her bills in return for services on the modernistic bed. We had some formal cat conversation with the cat's owner before she mewed up her property in a basket. And that was that.

It was dark as we waited for the train which would take us to the Polis. Our Greek nerves had begun playing up at the prospect of Byzantium. Sinister shadows streaked the platform as the train steamed in. At least we had no difficulty finding places. Anti-climax again. Till at Alexandroupolis an American slammed into our compartment, sweating and disturbed. He had become convinced there was a plot to murder him in the hotel where he was staying, and so, had fled in the middle of the night. At once sinister shadows recurred in the badly lit corridors and compartments of the Stamboul Express. Pasty, waxwork faces, but never that of a star in toque and eye-veil, peered at us through the glass.

Once more, anti-climax. During the grey tedium of dawn there was an argument between a customs official and a Greek lady bringing a vacuum cleaner to a relative in the City. The hose was uncoiled endlessly, the plastic examined to an accompaniment of soprano shrieks and bass rumbles. I don't know how the deadlock was resolved. We had begun the approach to the dismal fringes of Byzantium.

In Constantinople (we could hardly bring ourselves to refer to it as Istanbul in spite of the enthusiasm of the British and Australians) we put up at the Silk Palace Hotel recommended by Manoly's father, I don't know why, but he had his unexplained connections, and in his youth, the Romanian mistress with whom he decamped from Alexandria to Chicago. The Silk Palace had the doubtful advantage of being close to the railway station and the brothels. Outside our ground-floor window was a street urinal reeking of ammonia. We were saved from its fumes by a Roman housekeeper who had heard that Romans were checking in. (Constantinople Greeks have always referred to themselves as Romans.) Thanks to this good woman we were transferred immediately to a sweeter side of the Silk Palace, where in spite of the well-meaning housekeeper, the plumbing in our bathroom did not work, and the lift which should have carried us up to our room was permanently out of order. (In our first days in the Polis we learned that most facilities were out of order, or else, closed.) But even so, we were grateful for the concern of our fellow Roman, the housekeeper at the Silk Palace.

Constantinople – from which the grandeur and the gold leaf have peeled, leaving the sludge that is Istanbul, the grey-skinned Turks, the heavy, teeming herds infesting the streets and crowding the buses . . . As practically no average Turk seemed to speak anything but the Turkish language, of which we had about six words, we took to listening for sounds of Greek whenever we were lost or confused. Our fellow Romans always welcomed us as members of what was as much a secret society as an exposed minority. Living from day to day under threat of expulsion, these Greeks were the warm pulse in an otherwise insensitive city.

Numbed by traipsing through squalid streets, dazed by the opulence of bejewelled Ottoman artefacts in museum and palace, we found relief amongst the mosaics of St Saviour in the Fields (Kariye Cami) and in the calm immensity of St Sophia, noblest of churches, nobler even than that other temple the Parthenon. Its spirit has survived Islam's attempts to disperse it by hanging beneath its dome plaques painted the Prophet's green with gold inscriptions from the Koran. This great church

166

is the embodiment of an ideal, none of the finicky Gothic soaring and aspiring towards Heaven, but a balanced statement of conviction that the spirit is here around us on earth. Paradoxically, some of the nobler mosques are inspired by this envied and resented church. We should get the message, but it is difficult when blood has flowed and continues flowing.

We returned several times to Ayia Sophia. Closed doors on some of those occasions only kept us out physically, and I shall continue returning in memory to this supreme church which I first entered as a child through a crude colour photograph in my parents' house in Sydney.

Our other great experience of Byzantium was the Bosphorus, a silken waterway down which we made several voyages. Here at last Byzantine and Ottoman seem to have come to an agreement, old Turkish houses and the ruins of Byzantine walls and gateways reflected in the same placid water. There were small, civilised restaurants serving exquisite Greco-Turkish dishes. Perhaps the voice of the Roman tended to predominate. We felt at home – till the ghost of that English governess marching along the towpath at dusk during a previous century looked down and found herself face to face with a head floating past in a basket.

Into Anatolia

Worse and worse for the soul of a martyred Greek and anyone who has suffered through close association. The bus we took for Smyrna was crammed. It was perhaps less rackety than the Greek buses, and for that reason, more athletic, inclined to leap hairpin bends, and skid along the edge of a precipice. Of course the Greek buses do it too, but sight of the driver's patroness, the Panayia, swaying against the windscreen, strengthens the nerves of a lazy believer whose faith is buried in his unconscious. On the road to Smyrna we depended solely on fate and fallible human skill.

All the way down through Asia Minor our driver shoved records on his player, often the same one again and again, at full volume. A hairpin bend never stopped him changing the disc or persuaded him it might be wiser to keep his hands on the wheel. A Turkish officer draped himself on our headrests trying out his limited English on our necks. At Bursa where we paused he insisted on paying for our food and coffee. We were almost stunned by now, my Greek alter ego more than I, what with the Turkish officer's thick-skinned, though well-meant attempts at friendship, and the blast of the driver's favourite disc, 'Menexé giöz aman aman . . . ' ('Blue eyes oh! oh! . . . ')

Landscapes of a peerless beauty inevitably evoked the past — in particular that hilltop village rising out of the distant plain, dust-haze, dream, the depths of unconscious experience. I nudged M. He wouldn't look. He was sulking. Some way on, after wrestling with the map and the guide-book, I couldn't help sounding sardonic. 'That was *it* — Nicaea — if you had only looked . . . !' I too, was sulking by then, but my sulk had a sharper edge to it. So we torment each other round the world and back during the lifetime we could not share with anybody else.

We arrived at Smyrna after dark, and on shaking off the Turkish officer, who stuck like gum, found our hotel, a modern concrete cube in the motel style. There was nothing wrong about it, only that this should have been Smyrna, not Izmir, and so it continued except for rare intervals in the days which followed.

Without the associations and ghosts which had not been exorcised, our attitude might have been that of our friend A.S., who lived there for some time, teaching 'dazzling young Turks' and enjoying all the advantages. It was the end of our friendship with A. We had to put him wise to houses razed, family, friends, servants escaping with little more than their lives, and the Archbishop, his eyes put out, crucified on his church door.

For us, flames, smoke, blood from the sack of Smyrna, still polluted the air of modern, aseptic Izmir.

We strolled along the Prokymaia where George Lascaris and his sister Polymnia had been walked as children by their foreign governesses. Most Greeks cannot help but stroll, it is something to do with their legs and the nearness of the Near East, while the majority of Anglo-Saxons, especially Withycombes, march. So there is always a distance between us when Manoly and I go for a walk. No one else would see the invisible threads connecting us; it looks as though we have had a row (sometimes we have.) In Smyrna the gap probably widened as we strolled or marched beside the Gulf, its waters drenched with the tones of drowning stocks. Eastern Mediterranean waters create this illusion. The approach to most Greek islands lying close to the Asia Minor coast swims at evening not only with the colours, but a nostalgia which quenched stocks evoke. From the steamer's deck I experience an unhoped-for inner harmony which the emptiness and boredom of provincial life disperse soon after landing, and I long to catch the next boat away.

In spite of the ghosts, we did in fact enjoy several days of simple pleasure in Smyrna thanks to the members of another and different secret society. These were Turks repatriated to their homeland after what remains for Greeks the Catastrophe of 1922. In every instance those we met were humble men, taxi drivers, shopkeepers, or waiters, who had spent their formative years in Crete. Their first language was Greek. They loved to speak it, in confidential tones, with these strangers who had suddenly come among them. Turkish repatriates from Crete, like Greek refugees from Asia Minor, live for the most part in the past. They treated us with the kindliness one owes fellow sufferers. From our contacts with the 'Romans' of Constantinople and the Turco-Cretans of Smyrna, I understood how the early Christians must have sustained one another under persecution, like Jewish minorities and homosexuals in the present day.

Walking through the Jewish quarter of Izmir we had the impression at least that its inhabitants were leading tranquil lives. But you had to wonder for how long. As we loitered in the heavy sunlight, sounds and smells, glimpses of matted, black-haired Jewesses and grizzled artisans brought back Alexandrian streets, including our own Safia Zaghloul, before the Greeks and Jews were driven out.

When the time came to return to Greece it was still an escape from Smyrna, though without the blood and flames of the past; it was a matter of placating a leaden Turkish bureaucracy, which is difficult enough. Every receipt, every scrap of waste paper, should be kept in a special wallet. I could not account for one of the travellers' cheques I had cashed in Constantinople. Wherever we went in Smyrna trying to convince the authorities I was not engaged in some dark attempt to diddle the economy, we met with stony expressions on the faces of those who had the power to sanction our embarkation at Chesme. As we sweated through the streets of what was by now only Izmir, running from official to consul, and back to official, we lived through one of those nightmares blown up from a trivial incident into a surreal situation of international significance, which we still go through in actual life and sleep, except that in actuality, one of us can wake the other up by giving a shout from across the landing.

No shout would have worked in Izmir. Then, suddenly, I found the missing receipt rolled into a ball at the bottom of a raincoat pocket along with all the other trash, shreds of paper, lengths of string, the bus and theatre tickets it has been my vice to accumulate from childhood on. (From finding a dog-eared prayer-book amongst the childish rubbish in my pocket, my mother said for a time, 'He'll probably go into the church.')

At Chesme again we were sustained by members of our secret society – gentle, civilised Turco-Cretans or Greco-Turks, who fed and talked to us while Turkish police bullied us into paying their unofficial fee before they would allow us to stagger aboard the caique 'Aphrodite' and sail for Chios. So we 'bade farewell to glam'rous Turkey', ending our expedition into a far too personal Byzantine and Anatolian past.

Islands (1)

Islands become less attractive as I grow older, perhaps because in an over-populated world there are few that are uninhabited, or because in my middle age I stopped being afraid of myself and others and became more gregarious. (This does not prevent me wanting to avoid human beings who pollute islands.) But in my undergraduate days I would have given anything to be able to withdraw to some rock in the Scillies. I have not dared venture to the Scillies since, for fear of what I might find. Fraser Island off the Queensland coast is one of the few to retain their magic thanks to watchdogs like John Sinclair guarding it against destruction. I first went to Fraser Island after Sidney Nolan gave me the story of Eliza Fraser and the wreck of 'Stirling Castle'. I went there on my own and began the novel *A Fringe of Leaves*, but gave up on deciding Australian writers should deal with the Twentieth Century. Years later Manoly and I went to the island together and explored it more thoroughly. From two visits and a certain amount of necessary research, it became part of my life, and the novel I wrote as painful and sensual a situation as one I might have lived through personally whether as Ellen Roxburgh or Jack Chance. I shall not risk desiccating the novel by going into it in detail, just as in speaking of Greek islands I am not setting out to compile a guide complete with archaeological sites, dates, distances, and heights. What I shall try to show is how they add to this self-portrait I have undertaken, and the most important relationship of my life.

My first Greek island was Aiyina during the year I spent in Greece with the RAF at the end of the Civil War and the German Occupation. Manoly's regiment at this time was engaged in mopping-up operations in the Aegean, landing on islands still occupied by German garrisons. We had managed leave together during this blazing summer. We set off for Aiyina because of its accessibility. With us came Elly, M.'s youngest sister, and her second child Epaminondas (Notis). We sailed from Piraeus in one of the caiques which had survived the war, the very simplest of vessels, but for this reason we were brought closer to sea and islands. We sat on deck amongst dour peasants with the skins of armadillos, a finger or thumb sometimes missing, even a whole hand, and their kerchiefed women, vomiting almost before we had left Piraeus, crossing themselves as the swell took us or waves hit us in the face. Notis, a cherub off a palace ceiling, had never experienced the sea and was upset by what he was undergoing; he began to cry, not sure which end of him would give vent to the disturbance inside.

For me it was the first stage of an odyssey into a Greece I knew from literature and target maps, but which I now hoped would continue throughout my actual life. I had been told so often it can't possibly last, and my experience up till then confirmed that it doesn't. It did however.

After much pitching, rolling, and supplications from the passengers, the stubborn caique landed us on Aiyina, a long flat treeless, rather prosaic stretch of land rising at one point to a volcano, its classic cone extinct, anyway till next time. As the inns and other lodgings at the port had no appeal, we trudged a little way along the coast to where, we were told, we might find rooms. We found them attached to a *kendron** where we ate our meals, chiefly fish, beside the sea. The rooms themselves were in one of those cubist island houses surrounded by a pistachio plantation, its roots reaching down through hot flat sand to the sea water on which these trees seem to flourish. The proprietor, the Strategos, was helped by his two spinster sisters, Science and Convocation. The lavatory was a Greek urn which Science used to bail out every morning with a big iron spoon.

In the whitewashed house we lay sweltering behind shutters, keeping the mosquito at bay, but content. What might only be a pause in the endless upheaval we had learnt to expect, disguised itself as perpetual happiness. We did nothing beyond read, bathe in a lukewarm sea, eat real food under a brushwood canopy after years of messing, army- and air-force style. We also strolled alongside the placid straits, to which colour returned in the cool of the evening, and on the other hand, fields bleached yellow by heat and drought. We had no desire to visit the temple or climb the volcano, no sights or sites – though once we walked to the house where Capo d'Istria* had lived, in a cool, almost dank, orange grove along the coast.

It was while on Aiyina that we read in one of the rare newspapers we brought back from the port that the war with Japan had ended. It was too strange to believe – first the Germans, now the Japs. In the throes of our first relieved and exhausted reactions, we had little thought for how the ending had been brought about. That was for later. The future stretched ahead of us, purified and perfect, as it does apparently at the moment when any war ends, like the last shot in a bad film. All three of us accepted it – Elly after the horrors of a German occupation, hunger, humiliation, a child delivered on the kitchen table, Manoly and I thankful to relinquish the minor parts we had played in the services, and perhaps some of the subterfuges we were forced to adopt to disguise a relationship which was all-important to us. We were modestly if incredulously happy at this happy ending. What we could not have accepted at the time, even less acceptable than those bombs on Japan which only remotely concerned us, was any inkling that Notis, the cherub splashing in the white shallows of the Aegean, might grow into a bearded scientist, work in Germany, and marry a German. So we congratulated ourselves, and drank to the future under the brushwood canopy as we ate an excellent *tsipoura*,* not realising that the future has seven veils.

Difficult to remember in which order we visited the Greek islands. I see them as a patchwork, or collage, or progressive disillusionment, not only with Greece but with post-War life. Fortunately, if our relationship was ever rocked by experiences in our travels it always recovered its balance.

Crete turned out to be a bit of a trial. Although some years after the War, Herakleion was still the shambles to which it had been reduced. Everywhere oriental apathy and squalor. The most abysmal food submerged in oil, blowflies haunting the pyramids of donner kebab, lavatory chains which wouldn't pull, a stench of sewage. On the other hand sculptures of superhuman perfection: those figurines of Apollo, forerunner of Christ and many humble peasants, the lamb draped across godlike shoulders.

We did the sites, but this is not a guidebook. At Phaestos we met the bore who had known celebrities, and danced, even slept (?) with famous beauties.

On the road to Ayia Triada we passed a little dusty peasant boy piping to his flock on a reed.

Then, Ayia Triada itself, and the ruins of a villa where one could have moved in to enjoy civilised living at any point in time before the Germans parachuted down on the island and the English started developing it.

We leaped by bus from Herakleion to Canea with all the usual incidents from close shaves on the precipice to women spraying our necks as they vomited into paper bags. At Retimo, which I had known intimately as a target map, we rested on rush-bottomed chairs to a clashing of *komboloyia* dangling from the idle hands of bull-faced Cretans. After we had passed the turning, we realised to our shame we had not taken notice of the sign pointing to El Greco's village. Canea presented a more refined shambles than Herakleion, a touch of Eastbourne about its boarding houses. But we were followed by children shouting 'Cherman! Cherman!' It happens to me throughout Greece. Is it because my attempts to speak Greek are more persistent, or more convincing than the average British traveller's? (Perhaps not, for I still have the humiliating memory of the Theban pilgrim to the Grand Lavra who took me for a Brazilian because of my pronunciation of 'Australia'.) It might be my eyes, not exactly cold I like to believe, but I know they tend to resist seduction by effusive strangers. Poor Manoly on these occasions can only appear the collaborator. Though who, among the Greeks, as time goes on, has not collaborated with Germans? There has been a succession of German monarchs ending with Queen Frederika, Kaiser Wilhelm's grand-daughter who graduated to an ashram and a guru.

During the war I was one of those representing Air Force Intelligence in an operations room on Cyprus while an attempt was made at recovering the island of Cos. The operation ended in disaster, the landing party taken prisoner or escaping into Turkey by swimming the narrow straits separating island from mainland. The crucial night was a real nightmare as we tried to get in touch with senior officers lost in the nightclubs of Nicosia. Any such business throve during the war through the Hungarian and Romanian 'cabaret artistes' stranded in Cyprus. Far from cramping their style the British administrators saw these women as a means of entertaining the troops. One enterprising Hungarian bawd became the most influential figure on the island, and would explain to officers she was showing over the villa she had built, 'There, you see, every brick a prick . . .'

Long before exploring the islands of the Dodecanese in the round, I knew them by heart from target maps and pre-War intelligence reports. I could have led a conducted tour through many of those islands. I knew where the harbours were, the gun emplacements, and the convent of Italian nuns which fugitives were advised to avoid. Years later, while I was still in the grip of a passionate love affair, not so much with Greece as the idea of it, my introduction to actuality was a disconcerting, not yet completely disillusioning, but deflating experience, one in which Manoly shared.

Cos when we went there had all the sick signs and smells of extreme poverty. Even if in the main town there were left-overs from Italian colonisation such as examples of the admirably practical serial architecture I had seen in Libya, at the other end of the island children with disproportionate heads and bodies, like the figures in a Dickerson* drawing, sprang out of holes in the earth to beg. With all this, the serenity of the Temple of Aesculapius was wasted on me one glistening morning, and the great tree propped on its crutches a gangrenous sight. We were glad to escape, whether to better or worse.

We had come from better. In those days Rhodes had not yet been quite rotted by film stars and tourism. If you got away from the hurlyburly of the town, cruises arriving and departing, the Colossus and the Knights of St John offering scenarios for Anthony Quinn, you could still catch glimpses of island life. Again I enjoyed the game of converting target symbols into actuality, or indulging the imagination I had been forced to restrain during the war. Now, what had been no more than the squiggle of a line plotted by a mapping pen was shown to be a dry river-bed with pink oleander frothing between its white boulders. We drove on. The second image I have brought away from Rhodes is from the church at Lindos: its pebble floor not only an epiphany, but a memorial to the anonymous craftsmen who tapped the stones into place, and the villagers who have worn them down in worshipping. My third unimportant, but personally significant memory of the historically important island of Rhodes survives from the same village of Lindos: it is that of a woman reclining on the rampart of her house, who invited us inside to look at her ancient Rhodian plates, for no reason other than those of pride and friendliness. I see her still in the late afternoon light, the long, trailing, rose-coloured scarf, the line of an indolent Circassian thigh, on the battlement across from the Asia Minor coast.

Little queasy steamers carried us from Rhodes to Cos, from Cos to Kalymnos. We crouched on deck clinging to a brass rail as time and again the vessel threatened to turn turtle; better drenched by spray than asphyxiated by fug and the stench of vomit in the minute saloon below the water-line.

Sailing into the calm of Kalymnos at evening, its houses built in tiers above the harbour and washed in the colours of a bunch of stocks, disposed us for romantic reactions. Reality gave us a melancholy island empty of life, its women waiting for their men to return from sponge-diving in the Red Sea. We found a room in the house of a family with social pretensions and English connections preparing to leave their island the moment they were rid of their untimely lodgers. All night shutters banged in the gritty, half-empty house, which belonged to the owners by deed, you felt, not affection.

Driven round this barren island next morning, we stopped the taxi at a point where we were looking down from a great height on a green plain at the head of one of those blinding Aegean inlets. Like travellers in the desert, we could not wait to push on to this oasis, but were told by our driver it could only be reached by sea. So we resigned ourselves; in Greece the enchanted distance is often destroyed by close acquaintance.

We spent one more night in the house of banging shutters while the Anglo-Greeks packed. All of us decamped, not without relief, the following morning.

Island-hopping is an addiction, as remote from tourism as hard drugs from aspirin. Whether any of my fellow addicts catch themselves resenting their vice, as I do, I've never discovered. Perhaps we are too ashamed to admit. During our visits to Greece in the post-War years, I've been driven by my addiction for reasons difficult to clarify. The design or purpose of the crown to which I must add yet another jewel has seemed of less importance than satisfying my lust to climb another ridge, to persuade another rusty lock to give me access to cold candle-wax, a worm-eaten iconostasis, and the faces of saints, blotched, pimpled, scrofulous, half devoured by time and damp. Yet my lust has never been lacking in love unless frustrated by a lock which resists picking, and some peasant woman, keeper of the key and the candles, gone to work in the fields or to harvest the chestnuts, taking the key with her.

Possessed by what I shall dare call my lusting spirit, and Manoly's Orthodox devotion, we made our pilgrimages to Patmos, Samos, Chios, Lesbos, in what order I can't remember. Chios stands apart from the others in this group through memories of our escape from Turkey in the 'Aphrodite'. Chios should have been paradise. In fact it is a lump of pumice rising from the sea, its port a boring one even by island standards. We found an antiquated hotel, more like a block of flats, its rooms furnished with iron bedsteads and white cotton counterpanes grouped round a stair-well hung with ferns and other indoor plants. Some kind of drama should have been taking place on that stair-well and the honeycomb of rooms clustered round it. In the beginning we spent hours on the white bedspreads sleeping off the Turkish experience. When we felt we had resumed our normal shape and more or less a stable frame of mind, we went out to explore the town. It was difficult to find food in Chios, all along the quay jam shops, with never a sign of a taverna, unless farther in, an establishment providing lukewarm oily stews for anyone who dared risk the conse-quences. In the evening as we sat on the quay, drinking ouzo or a coffee, while spooning up some of the inevitable jam, the bourgeoisie strolled up and down, island style, the ladies armed with handbags, no doubt planning next season's Athenian handbag and their costume for the evening stroll.

More evocative were the walled gardens of one or two comparatively large houses surrounded by trees and moist-looking shrubs. The houses belonged to rich Chiot merchants who had made their fortunes in Liverpool and Alexandria and mostly lived away. These island houses reminded me of similar ones surrounded by walls and dark gardens in the Alexandrian suburb of Schutz which, together with Manoly's reminiscences of childhood, gave me the story *Being Kind to Titina*. From Chios too, the incident of the gipsy in *A Glass of Tea*, which actually happened to M. when he landed on the island with his regiment as Greece was recovered from the Germans.

We made several expeditions in a broken-down taxi along the rutted roads of this island which seemed more than ever a lump of grey, dusty pumice. We visited two memorable villages on a day spent mostly in mid-air between the taxi's seat and roof. Mesta, the first of these villages, has a square with frescoes worthy of Picasso, the second, Pyrrghi, had been built as a fortress. Its high-walled, narrow streets leading inward to a sumptuous church at its heart are painted with what look like shafts of golden light. A big red pointer barked a warning as we approached and loped off to rally the garrison against a possible invasion by Turks.

Planted throughout this part of the island are groves of mastic trees, gnarled and dusty to match a grey soil and pitted hillsides. Much of Chios looks extinct. Apathy has settled where passions erupted in the past; blood flowed without acting as a fertiliser.

Samos is one of the sadder islands, more wooded than some, but dank, its monasteries depleted, their frescoes damply mouldering. One establishment housed a monk and a nun, in what relationship we didn't gather, but hoped it was a happy one. The old nun looked placid enough in her vegetable patch, but you felt the monk could have thrown a few midnight tantrums.

At evening in the port of Vathy the bourgeoisie strolled up and down relentlessly, married couples linked, students (girls on one side, boys on the other) in unbroken chains beside the sea. The word 'ennui' might have been coined in connection with the evening stroll in island ports and the provincial towns of mainland Greece. Perhaps a suicide we knew to have taken place on Samos injected an extra melancholy into its rituals and influenced our attitude to what we saw.

At the tourist hotel where we were staying we caught sight of Henry Miller's 'Colossus of Maroussi', a vast, dropsical, sulking figure awaiting somebody, some event, or nothing, seated in the foyer in an armchair far too small for him. (Who doesn't sulk in a Greek tourist hotel? There is always the plumbing, the ration of Nescafé, the missing laundry, or simply the general ennui of island or province.) It turned out the flaccid Colossus was awaiting his wife, a lady who screeched like a peacock at the management, and had boiling water gushing in no time from bathroom taps. I'm prepared to think it was Mrs Katsimbalis, not the Colossus, who roused the Athenian cocks into action that memorable Athenian dawn described by Miller.

We were carried from Samos to Patmos in another tormented vessel, surrounded on deck by the usual complement of peasants with despairing kids, lambs with the knife already at their throats, bunches of hens, and an array of pungent little cheeses. This time the peasant vomit was joined by that of a party of theology students with the result that the cheeses were at least sprayed with sanctity.

We had great expectations of Patmos, what with the Cave of the Apocalypse, and the monastery founded by the Lascaris saint, Christodoulos. Our expectations were to some extent answered. Patmos was (anyway, then) an icon come to life, the miracle brought about by a conjuring between blue of sky and dun-coloured, rock-strewn earth. The season helped: branches were putting out their first, stylised leaves, rare blades of grass were shooting, florets trembling in clumps between the stones — all as the skeleton of Orthodox tradition demands. If it seemed probable that the Prophet Elias haunted his chapel, nobody would climb the peak to find out. On the saint's feast day a priest was reputed to pay his respects, but from what I had seen of Orthodox priests, I doubted.

At this period the majority of tourists landing on Patmos

were immediately seated on donkeys and led bouncing up the mountainside, a grotesque cavalcade, ladies with skirts hitched above their thighs, either sex clinging to the pommels of unyielding wooden saddles, to be delivered at the Evangelist's cave and the great monastery higher up. After doing their duty they were hustled down helter-skelter to resume their cruise. Peace returned to the island, and grace of a kind.

Our first morning we climbed to the Cave of the Apocalypse. Natural splendours made the cave itself seem tawdry, reminiscent of the booths and sideshows at fairs and circuses. At the back of my mind opened the tent at Rushcutters Bay, Sydney, where as a child I witnessed an act of creaking levitation, during which my soul was struck with terror when a young girl in the audience fell to the ground in an epileptic fit. The smells of overheated canvas, sweat, and face powder in an Australian summer, were not so far removed from those of burning candles, sweating rock, armpits, and sanctity I found at one of the sources of Christian faith. I did not truly believe until we came out again into a slashing sunlight. I was shriven as I stooped to pick a few sprigs of wild stock growing beside the path. One of these I still come across, pressed between the pages of my disintegrating address book.

On Patmos we met the corpse of the aged peasant woman they were carrying through the streets on her bier. In her best dress and apron, she would have passed for a figure asleep on a day-bed had it not been for the emotion of the mourners, and her bearers' staccato movements which tossed her from side to side like a lifesize doll too loosely filled. Plaited from girlish hair, probably at the time of her wedding, pigtails rustled like snakes along her neck, from under the kerchief which hid her grizzled scalp. The old leather slit of a mouth, sewn up callously by death, was intent on guarding its secrets as closely as a Greek village will allow.

It was on Patmos that Hurtle Duffield and Hero Pavloussi experienced a similar shock on being faced with an old woman's corpse. Hurtle noticed the man pissing against the wind on their way up to the monastery, an image which so disgusted an English lady living in Vienna she wrote to tell me she could not understand how she had ever admired my books. As Hurtle and

Hero we also visited the convent on the other side of the island and were entertained by the abbess-odalisque overlooking the Asia Minor coast.

What I continue to remember most vividly is the house on Patmos we thought of buying. We came across it in the maze of streets somewhere south of the monastery. The key was produced by an agent of the owners. There was a crude wall round what was not so much a large garden as a neglected almond grove. It had its own well. The house was too large by practical standards, but ideal for one who likes to walk from room to room while stuck for a word. What I remember most about the house was that its emptiness was waiting to be peopled by characters out of Chekov's plays — a first warning. The Russian influence would have been introduced naturally enough by the Greek from Odessa who had decorated the rooms with their sombre tones, dull reds and browns, dados painted in the key pattern, or classic urns. Almond branches rubbing on the windows created inside a melancholy which seduced that part of me which is no-hoper and masochist. At the same time blunt doubts began assaulting my entranced conscious: earthquakes had cracked the house and garden walls; in amongst the almond blossom reared the stone equivalent of an Australian dunny, a sculpture of petrified shit on its seat; and most pertinent of all warnings, a glimpse of Turkey across the straits.

Back in Athens, in spite of the fortune the owners were asking for their dilapidated property, my fantasies continued flickering round the house. I furnished it. I traipsed down to the harbour behind the donkey we kept to carry our supplies up the mountain. The light in the rooms of the re-decorated house encouraged the creative impulse. In the kitchen some unquenchably amiable village Tassia or Stavroula, dedicated to the whims of her employers, stood fanning the charcoal, or watching over pots from which rose smells of onion, garlic, and tomato simmering in fruity olive oil. The unruffled hosts, awaiting unpunctual guests from Athens, lead them out to the courtyard at last, where we sit beneath almond blossom, drinking our wine, eating our simple peasant meal, making civilised conversation.

Perfection — till storms arise offshore, and in the house,

where guests have been forced to outstay their welcome, malice
abounds. Food runs short. The post-card donkey waits in vain at
the harbour to carry fresh supplies up the mountain. In the
walled garden, rain and the splatter of almond leaves fail to
drown the sounds of diarrhoea from the stone dunny.

It was a relief when our dream of Patmos petered out finally.
If we had forced our fantasies as far as actuality, schizophrenia
and bankruptcy would have got us as we jetted between the
hemispheres. Still, I hanker, my regrets niggle, and occasionally my bower-bird's eye returns to carry off a detail from the
Patmos house to decorate one of my own fictions.

A Detour

I sometimes wonder how much of my life has been mistaken.
Should I have been able to control it had I been offered the
opportunity to do so? Or was I able and chose not to? I never
re-read my books once I have corrected the proof, but if for some
specific reason I have to open one and glance at a paragraph or
two, I am struck by an element which must have got into them
while I was under hypnosis. On one level certainly, there is a
recognisable collage of personal experience, on another, little of
the self I know. This unknown is the man the interviewers, the
visiting professors, the thesis writers expect to find, and because
I am unable to produce him I have given up receiving them. I
don't want to pretend to be me, as the poet Philip Larkin has
said of himself in similar situations. The masks I put on in my
fictions are very different from those which strangers try to force
on me, or to use another metaphor, the characters of whom I am
composed cannot include those not yet revealed to me. At the
age of sixty-nine I am still embarking on voyages of exploration
which I hope may lead to discovery.

Of course there are individuals, often very close, including
the one who is closest, who consider they know me better than I
know myself. These are the most deluded of all. It saddens me
that those I love, have to make do with superficial acquaintanceship when I would have them enjoy complete union. But I am

this black, bubbling pool. I am also this leaf rustling in the early light on the upper terrace of our garden. In the eyes of God, the Eye, or whatever supernatural power, I am probably pretty average crap, which will in time help fertilise the earth. The books I have spent years writing will be burnt in some universal, or perhaps only national, holocaust.

That is what I think tonight; no doubt I shall see differently in the morning, and as differently on every other morning I am fated to live through.

Islands (2)

At the time when we went to Lesbos the island had little appeal for tourists, perhaps because of its size. For the same reason, Euboea has been neglected because those foreigners who want to possess, and who incidentally destroy an island, see that vast tract of mountainous country as part of the mainland since it is close enough to be joined to it by an insignificant bridge. Lesbos is smaller than Euboea, but has the advantage of being farther from Athens. If the Sapphic legend attracts a dedicated few, it can hardly be used to promote a tourist industry, and indeed the town of Mytilene at one time refused to let a room to a couple of women, while accepting a pair of men. Perhaps in revenge, lesbians who had been turned away and forced to spend their nights on beds of pricking pine-needles, created the myth that Mytilene is famous for its olives and its buggers. I don't remember coming across either a lesbian or a bugger during our stay in the flat, characterless town of Mytilene, though the olives were large, black, and juicy.

Lesbos revealed itself as an island of several faces. Unlike Skyros, another fairly large, but shy, or downright secretive island, Lesbos comes out in all its colours without over-much wooing.

Not far from Mytilene, along a continuation of the flat, relaxing coast, there is a string of substantial houses washed in faded ochreous to terracotta tones. Rich Anatolian families of the past, among them the Lascaris grandmother with her daughters, used to rent these houses, to enjoy a change of scene and take the waters. Today the houses are silent, shuttered, in varying stages of dilapidation: no sign of maids beating mats, plumping out pillows and mattresses, airing sheets. All maids, it seems, have defected to Athens, or worse, Australia and South Africa.

Molyvos, a pretty village perched above the sea on the north coast, has become the victim of international intellectuals and artists. They return year after year, to paint, write, and discuss the reasons why their marriages should break up, even type poems to the marriage which has broken that morning. We ran away from Molyvos. A reminder of our visit is a striped cotton cloth such as peasant women use for a kerchief, which I bought in the village while our driver was digesting his lunch. I foresaw no practical use for it, but over the years this fading yellow Lesbian kerchief has been protecting my typewriter from dust.

Here and there across the island we found the vestiges of murals by that great painter Theophilos, the most impressive of them at a *kendron* under a giant plane tree not far from Mytilene itself. These vast, gnarled trees are usually the hub of village life. They are meeting places of peace and cool in the blaze of summer, of dank melancholy and neglect when winter takes over. The primitive figures of Theophilos are related to the popular puppet character Karagiosis;* all celebrate an age of heroism in the war against the Turk which resulted in Greek independence. Across the Aegean, on the same parallel as Lesbos, Theophilos crops up again. On the Pelion you find the same semi-fortress tower-houses, the same reminders of the fight against the Turk. But the neglected murals of Theophilos are elusive, because the keeper of the key is more likely to be visiting the relatives down in Volos, or harvesting apples or chestnuts on the mountain. It may, of course, be a peasant's stratagem of avoidance, to protect her eidola from strangers' eyes, as the monks of Mount Athos hold out against showing their treasure to pilgrims unless the latter possess granite determination.

Somewhere not far from the port of Mytilene was a mountain village given over to mild industry, in it a church where you felt faith was flourishing, the saints encrusting the iconostasis glow-ing with peculiar grace, and more unusual, preservation. What distinguishes anonymous peasant iconography from the work of the great Byzantine artists or a present-day master like Kontoglou I should find it difficult to explain. As one who is not an expert I go by my sensuous instincts when approaching such works of art. They seem to display a humility, a toughness, a direct, illuminating fervour. Or am I seeing in these icons what I would like, and often do see, in the peasants themselves?

I find the same qualities in what amounts to the almost extinct aristocracy of democratic Greece, where members of the great families became republican in the days of Venizelos. A new class arose round the German monarchy foisted on Greece by the Great Powers. The older, aristocratic families remained closer to the people than this other, new-rich bourgeoisie. Members of older families cultivate demotic speech often to a laughable degree, and sometimes embrace the Communist faith. It is not unusual for a sophisticated Athenian lady to pride herself on being a cook, and to ring in the postman at a dinner party, while her bourgeois counterpart, closer to the *précieuse ridicule*, works at a never-finished thesis.

On Lesbos there is a meandering inlet which at its closed end resembles a brackish lagoon choked with reeds. Here our taxi slowed down and eased past a procession of two limousines and a string of battered island jalopies. Our driver explained that a family of wealthy islanders had returned from the United States on a visit to their village. We could imagine that visit: the self-important airs of the expatriates, the emotional outbursts, the hospitality scraped together by poor households, their admiration and envy of those who had achieved longed-for prosperity and, basically, corruption. Heartache, joy, and resentment intermingled. I remember this scene when I see sad-eyed peasant grannies sitting proud but unhappy under their black kerchiefs in the family Holden in Centennial Park, Sydney.

The day we met the triumphal procession of expatriate Lesbians we had driven as far as Sigri in the north-west corner of this multifarious island. Sigri is as remote from the pretty airs of Molyvos as Wagner from Massenet. Only the chthonic gods or pre-Olympians could have inhabited such scenes of volcanic upheaval, and still haunt its aftermath of petrifaction, sour pasture, and contemporary poverty. Bumped along the road by our dislocated taxi, warned off by the landscape, we might have been cavorting on our last ride. First exalted, then irritable, one became desperate, moody, resigned – in other words, one went through the whole range of reactions roused by a relationship with Greece. Intermittent with exhaustion on this rackety journey one was fortified by a kind of sense-drunkenness, through which flickered the images of *pallikaria*,* peasant faces dressed as saints for the transcendental ball, and pasted on the leached-out evening sky, or tacked to hoary olive trunks, faded sepia snapshots of the Smyrna ladies engaged in literary pastimes on the terrace of their rented Lesbian house. Till finally, on arrival, the spell was broken. We washed off the paste of sweat and grit at the Hotel Sappho, and sat eating our oily meal beside the still waters of Mytilene.

Two Sea Pictures

On the voyage to Paros, Western ladies in oriental shifts, amulets, and an excessive number of silver rings, outnumber and also outvomit the kerchiefed peasant women. It is one of those Aegean days when a burnished sun and a glassy wind slash at each other above a rolling sea in post-card blue. Not far from us stands a German John the Baptist, his ropes of tow flung in all directions. Shivering inside the sleeveless red-fox jacket, he staggers repeatedly, but the classic sandals, laced to his shins and varicose veins, keep a hold on the runaway deck, while down from one leg of his cotton drawers, dangles a mauve yo-yo.

Most of our fellow passengers disembark at Ios, an island known in the old days only to real connoisseurs. Today as we approach the harbour we are faced with a square, three sides of it composed of hotels and tavernas. Inside the square revolves what appears to be a rehearsal for one of the hairier musicals.

At a slightly earlier date we were returning to Mykonos after a visit to Delos. On this occasion our fellow passengers were mostly American youths and German girls in separate parties, the Americans a mishmash of the races, the German *Mädels* golden-skinned, stout-limbed, with those pale blue eyes which can give the possessors a look of hypnotised fish. A storm was threatening as we embarked. Halfway to Mykonos a gale struck. Icy white shawls were flung across the deck of our labouring caique regardless of any human life irrelevantly clinging to its rails. Walls of solid water were erected at our bows' approach. Alternately we plunged into grey, seething moats at the foot of these watery fortifications. We were rolled first on one side, then the other. Saved from turning turtle only by a miracle, or the efforts of our grim, unshaven crew in sodden jerseys and denims rolled above hairy shanks, the most we could expect, if we escaped drowning, was to be washed up on the shores of Naxos, a substantial blue shadow to starboard. During all the wrenching of timbers, tendons, muscles, and swirling in the whirlpools of emotion to which we were being subjected, the American boys openly expressed their terror. They cried and hid their faces and clung together. Some of the stronger tried to wrap an arm round a weaker neighbour. While the German *Mädels* sat upright, their skins streaming, their fish-eyes more than ever hypnotised as they belted out folk songs in unison. We sailed into calm at last the other side of the Mykonos mole and put back the masks we wear in what is seen as normal life.

Islands – a Mixed Bunch

Nobody writing a book on the Aegean Islands would link those I'm about to include. They are dissimilar in character and from different groups, but each plays a particular role in my relationship with Greece and Manoly. Over and over, during these journeys and after, when M. tells me I hate Greece, I cannot explain my love. Again, in our more bitter, alcoholic arguments in the kitchen after the evening meal, when he tells me I hate *him*, I cannot prove that what I believe in most deeply, the novels for which my conscious self can't take full responsibility, our discomforting but exhilarating travels through Greece, our life together, its eruptions and rewards, my own clumsy wrestling with what I see as a religious faith — that all of this is what keeps me going.

Anyway, we sailed for Santorini, a volcano for the moment extinct. Though arrival is always the best part of the visit to any Greek island, the approach to Santorini scores over others from being in the grandest operatic tradition. As we glided inside the roadstead towards the glowering volcanic fragments known as the 'Burnt Ones' the sun had spotlit the high Santorini cliffs. Their strata of coloured lava reminded me of those glass domes filled with layers of coloured sands I had seen in my youth in souvenir shops on the Isle of Wight, and which I had always associated with the mantelpiece in boarding-house parlours. Perhaps it was the link with Ventnor which encouraged English eccentrics to settle on this Aegean island in the past. But that evening, as we shuffled impatiently amongst our baggage on the deck of the *vaporaki*,* the Santorini cliffs were shining with sheer operatic splendour. Dusk had begun weaving as we snoozed between the cliffs on the one hand and the dispiriting Burnt Ones on the other. Still lit by natural light, the cubist houses on the heights looked like a string of white vertebrae. With dusk increasing, the ash from the Kaimenes behind us seemed to rise in ghostly feelers out of the craters, till darkness set solid, and suddenly the cliffs above were hung with electric garlands, and clusters of light bloomed round the anchorage for which we were making.

We had arrived. Not yet the anti-climax. There was the scrimmage of lugging our baggage from steamer to motor-boat and from boat to quay, where a chorus of Verdian muleteers descended on a secondary chorus of soft strangers, the brigands mounting their prey on an assortment of unco-operative mules and driving them willy nilly up the *skalopatia*,* accompanied by shouts, protests, the smell of urine and plopping dung, and sound of hooves on the cobbles as we zigzagged up and up towards apotheosis on the illuminated heights.

For once we found a civilised hotel with plumbing which worked, linen sheets — perfect peace. Otherwise the minimum of food, as on most Greek islands. Tomatoes which grew straggling on fields as burnt and dusty as any Egyptian desert. Fish, when the hordes of German tourists hadn't devoured it before you got there. I can remember Kyria M., a native of the island, telling me in Alexandria how the quail migrating from Europe to Africa, would flop down exhausted on Santorini, and the inhabitants would run out and net them, to preserve in jars against leaner seasons. There was no evidence of preserved quail at the period when Manoly and I visited Santorini. It was some years after the last disastrous earthquake, when many of the island's more baroque architectural jewels had been dispatched. The cosmopolitan eccentrics were gone. Hippies stretched out in sleeping-bags on the cobbled paving at the entrance to the church. It was a period of tourism and decadence. Though when had the seeds been sown? I remember from talks with Kyria M. how she and her daughters used to escape the Egyptian summers by visiting relatives on Santorini. These were still the days of preserved quail. The women used to sit round gossiping and playing cards, or cooking the little there was to cook. Years later, as we wandered through Santorini ourselves, I could feel the sweating cards they dealt one another, hear the rustling of the thin gold Egyptian bracelets on the Alexandrian women's arms, and the laughter of bored, provocative girls. I remembered Loulla M. telling me in Alex in wartime, 'By October, you know, we feel destroyed. Nowadays there is no escape.' She laughed and shrugged. There was an air of exhausted, lubricious acquiescence about the young woman's movements and her tone of voice as we strolled along the sweating Alexandrian street.

None of the M.s' relatives remained by the time Manoly and I arrived on Santorini. There were only echoes. We visited a monastery where, amongst hanging baskets of lush-limbed, over-green, exhausted plants, an etiolated monk believed he had heard of our Alexandrians.

On Santorini we ran into our friend Fritz Tonnelle, an American expatriate, one of those refugees from the West who haunt the Near and Middle East. Another friend, Robert Liddell, calling on Fritz in Alexandria before the Second World War, found him reclining languid on a sofa, surrounded by books, vaguely scribbling. Asked what he was doing, Fritz replied, 'Ah'm translatin' Prewst . . . ' I don't believe anything came of his writing, but he also painted in a desultory way, to please himself, and some of his Greek landscapes convey the essence of Greece more subtly than Greek chauvinists and professional painters would like to hear me say. I associate Fritz with the ramparts of Santorini at sunset and its narrow, eventless streets, as well as the dank tank of his Athenian basement flat, window-less except where his own landscapes admit an outer world.

Thera is curiously lifeless for a volcano which history has shown to be no more than temporarily extinct. On the island's eastern coast tourists bathe from beaches of black, crumbled lava. There is the archaeological industry too, from which all the most important loot is carried off to Athens. From a terrace in the south you can look down on a sterile plain reminiscent of Egypt's Western Desert. Always these reminders of Egypt: the rustle of thin gold bracelets on ghostly Alexandrian arms, brown dust brought by winds from across the sea settles on your teeth, grit forces your eyes into a tighter, more protective squint. Perhaps to atone for the prevailing lethargy of the inhabitants between eruptions, the winds are alive, the sea aggressively so.

The inevitable storm arose as we expected to leave Santorini. The steamer could not put out from Crete. During these days we wore a track to the agent's office asking for news. We spent hours in a nearby *kapheneion* drinking cup after cup of muddy coffee and slugs of ouzo. Or I sat at the dressing-table in our hotel room writing a story, that saving occupation while wait-ing for the steamer to deliver you from a Greek island (this time I believe the story was *The Evening at Sissy Kamara's*). Has my writing always been in some sense a 'saving occupation', whether short stories on those Greek islands from which you do eventually get away, or interminable novels in the city which I shall never escape because we are linked by heritage and fate?

We were finally summoned from our room in Santorini at two o'clock on a morning of subsiding storm. The steamer was expected from Herakleion. We had been packed for days. We flung on our clothes, stubbing fingers on normally accommodating buttons, chasing the leaves of a suddenly flighty manuscript. We made for the assembly point, where the same Verdian chorus of brigands could be heard in full voice as it loaded baggage for the subsidiary chorus of soft strangers. Tightened girths made the mules arch their backs, inflate their bellies, and fart like machineguns at the jab from a knee. We were set at last for the descent by *skalopatia*, the same clopping and clacking, the scents of dung and urine, this time through the glaze and shiver of a too early unorthodox awakening. Arrived at the anchorage, we sat alternately in a fug from primuses in the unventilated *kendron*, and on a draughty quay, awaiting the arrival of the *vaporaki* as darkness ebbed and a bleak grey was insinuated between the extinct cliffs of Santorini and the ash dumps of the Burnt Ones.

The *vaporaki* anchored at seven. We stumbled towards the motor-boats which would ferry us out to the 'vessel'. I must have looked particularly ancient in my laden state and the grey dawn, for one of our fellow passengers, a young mulatto of grace and charm, speaking a cultivated English, insisted on carrying my bags. In transit on the motor-boat he broke spontaneously into song in a pure, unaffected voice. There was no time for more than some conversational banalities before we were swallowed up by the most abysmal *vaporaki* and submitted to one of our stormiest Aegean crossings. One of my virtues, a deplorable flaw to those lying prostrate around me, is that I am never seasick. Manoly almost never is, but on this occasion succumbed. We were seated in the shuddering saloon with its smells of oil and left-overs and the stains on a grubby tablecloth, wondering whether we could face a cup of coffee, when a stout *bourgeoise* passing through seized hold of a rail and rose on her points as though limbering up at ballet practice, then suddenly dropped her pose, darted at the galley doorway, and vomited into a swill bin. It was the cue for Manoly's exit. He lay low during the heaving of an interminable voyage. I saw very little of him, and only once caught sight of my young mulatto stretched on a bench almost entirely hidden by a blanket. One small patch of visible skin had turned a greenish yellow. I never heard his story, nor saw him again. There was the struggle of arrival at Piraeus, the fight for a taxi, descent into the flat air of metropolitan life, relief at being received back into the orderliness of the Polymeropoulos household.

Part of me hankers after domesticity and orderliness. There is also that face which looks towards the wild side. During our brief stay on Naxos (we were not washed up, but landed as planned) the one recoiled while the other exulted. Mythic from a distance, the port is a squalid little town, none of the whitewashed glister of so much island architecture, but smudges, smears, shit, and stale food in tepid oil when food is to be found. Naxos is a Roman Catholic island and Orthodox Greeks blame Popery for its squalor. Whether that is so, the moment the taxi we had hired to drive us round the island leapt into the interior, the stained veil was lifted. Naxos is large, wild, harsh, operatic (nothing of the effete R. Strauss in its orchestration). It was a long drive in a short time. On the rocky northern road, boys ran out to stone our windscreen. The driver stopped, cursed, and got out to retaliate with rocks on the little crop-headed boys. It always happened, he said, at the approach to this particular village. One felt he might have blamed it on Catholicism, though it was more probable the boys had heard their elders telling of skirmishes with Germans. In fact the driver's conversation was full of echoes from Hitler's War, tales of friendly raiders arriving by night at northern coves, and from the previous century, defence of the island from raids by the Turk. We finished our drive in darkness on the stoniest part of the whole road. The driver thought he had hit a partridge and got out to search for the corpse by light of his headlamps. He did not find it. The partridge must have escaped, as the Greeks invariably do from any of their predators.

I dissociate Naxos and Delos as far as possible from that passage to Mykonos during a storm with a gaggle of weeping American boys and jubilant German *Mädels*. I would also like to forget the hordes of tourists calling at Delos on their Aegean cruise, trampling the grasses, lumbering through spring flowers, gaping at every classical cliché in the book. It is their right, I know, and I shall stand accused once more. I would like to think that they too, can experience the lifting of the veil, but it never happens in the presence of guides or lecturers. Poetry resists academic pretension, just as the mystery of religious faith evaporates on contact with dogma.

Manoly and I once spent a night on Delos. We had the small primitive tourist pavilion, we had the whole island to ourselves. Walking amongst the ruins, past the broken columns and crouching lions, the heat rose in its last gasps, together with the perfume of wilting flowers, out of the waves of exhausted grass. There was a scuttling of lizards and a crying of lone birds as silence gathered palpably around us.

Back at the pavilion the 'Begum' served us our Spam, wartime style, and sat knitting at a slight distance, yearning to be changed into an Athenian *bourgeoise* in a claustrophobic unit. Most of the island women knit, no doubt as I write stories at hotel dressing-tables waiting for the steamer to arrive.

After finishing our Spam (as in our desert days, we could have eaten as much again) we stumbled up Mt Cynthos, behind its peak a moon rising in a periwinkle sky. We were dazzled, bemused, drugged by visions and scents. Divine dispensation had brought us together with the spirits of the place for a single night. We were led back finally and deposited on the army blankets and truckle beds of the concrete cube referred to as a tourist pavilion, through which the snores of the Begum and her curator-spouse cannoned and reverberated. Regardless of gross humans, the twangling, tinkling, rustling, the leathery croaking, all the night and sea sounds to be expected of such an island, interwove the other side of the concrete shell.

During the night we were glad of our rough blankets. We rose in a chilly dawn, stamped round the island under a pale sunrise, drank our ration of Nescafé, paid our bill, and waited for the boat to carry us back to Mykonos. Our awakening from enchantment had left us somewhat irritable.

Skyros, the largest and remotest of the Sporades, refused to reveal itself at first. Frustration dogged us. Stormy weather and unwilling boatmen prevented us visiting those parts of the island we most wanted to see. Roads hardly existed in the interior. On days of worst weather we were mewed up in a depressing tourist hotel on short and indifferent rations. The climb to the heights by cobbled *skalopatia*, past a kitsch statue of Rupert Brooke, had me broken-winded long before we reached the summit and the island's most important village. (It was at a time in my life when I felt that my scarred lungs were surely giving out.)

The village was quaint, cramped, and not particularly friendly once the inhabitants saw we were not the kind of tourists who are good for business. Some of their antique plates are of aesthetic interest; otherwise dwarf chairs are the most characteristic product of Skyros. When I first got to know Manoly I spent agonising hours through long Alexandrian afternoons in the Malliakas house perched on a dwarf Skyros chair, my knees almost under my chin, and carved knobs eating into my buttocks and kidneys: a martyrdom which turned out to be worth it.

But in the damp, cramped Skyros streets, surrounded by storms which refused to lift, I could not feel sentimental about the doll's-house furniture. We both became depressed and irritable, until forced by boredom, we struck out along paths which led to the interior. The island began wooing us, at first grudgingly, over stones, between thorny bushes, luring us down the friable banks of hidden gulches, then opening out into a dry white riverbed and an oleander forest in full pink bloom. There was a whitewashed chapel which two women, mother and daughter, had come to clean and decorate for their saint. The innocence of these two simple creatures was pure, touching, and is still revitalising to remember. On our Skyros walks I experienced revelations similar to those I could remember from exploration of the very different gullies scoring the flanks of Mount Wilson, and from a long ride through the Australian Alps on which I lost and found myself.

In the end we almost didn't want to leave Skyros for our next staging post. Skopelos is a more 'beautiful', obvious, sophisticated island, where Fritz Tonnelle turned up again. Obvious beauty, never-ending gossip of Athens and Alexandria, and the plumbing in the hotel where we put up, had me ready to leave as quick as possible.

Even more obvious than Skopelos, Skiathos had been developed as a retirement village for the well-to-do British middle-class, who discuss passionately in buses the difficulties they have found in what had been represented as a geriatric's paradise. Plumbing of course enters into the conversation.

For me the most vivid memory of Skiathos, apart from the Frenchwoman locked and screaming in the lavatory during the siesta, is the house of Papadiamantis. This bachelor public servant whose barbed-wire entanglements of style are liable to deter the reader, retired to his native island to die of writing and drink in his little claustrophobic house. His spirit haunted me as we climbed the narrow staircase and stood around in the minute rooms half-listening to the commentary of a worthy but superfluous guide telling us that the writer had been 'a friend to the bottle'. For God's sake — as though one didn't know instinctively! Good Papadiamantis, I share your rustiness of literary style, and understand your *Murderess* on whom you project your midnight self.

On a bland morning we left Skiathos. The same ferry was carrying a party of elderly English settlers bound for a mainland spree. You wondered how long they would accept exile from their natural neo-Tudor world, and hoped for their sakes that Papadiamantis and Greek plumbing would send them packing before complete senility set in.

Those tourists who are island-destroyers had not yet realised at the time of our visiting Euboea that it is not part of the mainland. Large hotels built on the coast beyond Chalkis had failed to attract and were already closing down. Outside the one or two still patronised, unhappy knots of pallid or lobster-coloured British stood disconsolately awaiting transport to take them on a promised outing.

We made our headquarters at Chalkis, where clouds of yellow smoke from a cement works effectively blanket its mythic associations. We stayed in what had once been a luxury hotel, now seedy, and smelling of overworked sewers. Even so we were

happy enough in our 'Twenty-ish room, battling each morning
for an extra envelope of Nescafé, and hanging our underpants
and socks to dry on the balcony above the view. I don't think we
quarrelled more than somewhat, though I remember being
depressed and edgy from coughing blood. To cough blood is
something to hide from even the closest of travelling companions. Though it had often happened, whether from the dust,
the jolting, rapid changes of temperature, or heavy, salt, island
air, I was cast down each time, and always felt I had to hide it
from one who is a victim of anxiety.

Our first afternoon we went in search of a dependable taxi-
driver, and struck a bargain with Margaritis, an amiable young
man whose wife was expecting her first child at any moment.
We were a bit apprehensive of the effect the birth might have on
our travels through Euboea, but Margaritis did not seem to
worry, and drove us the length and breadth of this splendid
island except for the day when the baby came.

We reached the convent of St John Kalyvitis after a drive
through a pine forest along an unfinished road. We bounced to a
halt in a cloud of dust, red for a change, and the prickling
silence of pine trees. The air around us was that mingling of cold
with hot; the light had a scintillating brilliance which one finds
only in the mountainous parts of Greece.

Not a Byzantinologist, I could not truly appreciate the
scabby murals in the chapel where some restoration from damp
and neglect had been undertaken through the persistence of the
abbess.

I soon realised the abbess was our reward for visiting the
convent. She came out to meet us, a tall, distinguished young
woman, the dark of her livery skin intensified by the black coif
surrounding the face. She led us on a stately visit of the chapel
while explaining with controlled passion her mission of restoration. One wondered what other passion was shored up behind
her scaffolding. There were no clues. She marshalled her three or
four subordinates with a dignity one would have expected of
some Byzantine aristocrat, and entertained us along with our
taxi-driver to coffee and cold spring-water. She lived on faith,

hope, and what else – and from where had she come? Nobody would have dared ask. Driving away from the convent, it occurred to me from knowing Voss-Laura to be myself, that the abbess was another aspect of Laura Trevelyan.

We continued along the road to Steni, a mountain village, alpine in character, with winding streets, cold air, cold water, plane-trees already autumnal in complexion, a church full of cold wax, and instead of the expected Byzantine mould, a clutter of framed oleographs. After driving higher up the mountain till threatened with losing ourselves by leaping into nowhere, we returned to Steni and bought a jar of mountain honey. The young woman who sold it allowed us to taste the various flavours, while her snotty child, daubed with honey, played round our ankles. Honey dripped everywhere in the concrete basement cube where the mother conducted what should have been her poetic business, but there were too many cans, and the snotty child, and I felt too ill from the drive up the mountain and the shock of alpine air.

Always churches. Manoly, the Orthodox traditionalist, accepts them as a matter of course. To me, all our travels in Greece have been pilgrimages, searches, whether ferreting out a nest of chapels hidden from the Turk amongst the elderbushes and sycamores, or waiting for an island steamer, or at the crossroads for a delayed connection, or in an abysmal bus station amongst the littered plastic, cigarette butts, and spittle, to be carried on towards the goal we only reach in diminished terms.

I am always hoping for the miracle. We are driven by Margaritis to Kymi and the spring which, he tells us, cures. Nobody has asked for a cure. We are all three rather jocular about it. We arrive at the spring, its water gushing out of a concrete wall. The pilgrims, or invalids, stand awaiting their turn to fill plastic containers, bottles, or to drink from picnic cups they have brought with them. We have nothing. I go up to the inevitable *kendron* to see whether I can borrow a glass. They are not willing. I return to the miraculous spring. My chest has been at its messiest. By now my tongue is hanging out for what amounted to a sacrament. I try catching the water in my hands.

A man approaches and offers to lend me his plastic cup. I look up gratefully and notice the diseased mouth making me this offer. He is too innocent to be daring me not to accept. I am the one whose lack of faith can never hope for a miracle. Even so, I accept the cup. I fill it. I drink from the pink plastic lip. I return it to my Samaritan, and Margaritis drives us back to the infernal Chalkis. As we are bumped around in the taxi, I sit rubbing my probably infected mouth with a balled-up handkerchief.

One church we found after fighting our way through long grass and suffocating elder-leaves was a little cobby creation of purest Byzantine form, like one of those old arthritic, aristocratic peasant women, brown and knotted, you see hobbling through the grass or vegetable patches adjoining chapels of this kind. Inside the building there was what you come to expect: mould, candlewax, often evidence of vandalism. This particular chapel, along a lane straggling from the Kymi road, I used as the church in the story *Fête Galante*, together with the mountain village of Steni, and the township of Nea Artaki, one of those neater, self-respecting Asia Minor communities to which we escaped one night from Chalkis. As we ate our meal at a taverna beside the sea, a lot, if not all of the story, erupted around us.

We were to have spent the last days of our stay on Euboea at the spa of Aedipsos, chiefly because of the associations it had for Manoly, and for myself from having lived at second-hand the lives of an adoptive family, boa-constricting its members for horrid creative purposes. I could not wait to reach this spa. From a distance in time I could smell its sulphurous waters. Languid aunts were taking the cure while indulging in intellectual flirtations with heavy-lidded gentlemen, their vices bridge and versifying. In my mind's eye they were still seated there under the plane-trees during pauses in the exhausting cure, twittering inspirational ladies of the protected class, and their costive men, successful at somebody else's expense, brooding under hats with curling brims, veined, arthritic hands encrusting the crooks of walking-sticks.

The drive to Aedipsos took us through the forests of the Noel-Baker pasha-dom. Noel-Baker, a British Levantine on the spot when Greece was freed from Turkish domination, bought his vast estates for next to nothing from Achmet Aga the departing governor. The Noel-Baker family (sometimes Baker-Noel in translation) have remained in possession till the present day, to the disgust of liberal-minded Greeks and British. Yet driving through the forests of this Levantine demesne, one has to admit unwillingly they probably would have been destroyed had it not been for the feudal vise. In the chief hamlet and administrative centre, there is a church housing the body of a saint. Bits of him have been carried off in spite of the ruling autocracy which, perhaps because more Breetish than Orthodox, does not worry all that much, while seeing fit to encourage sacred obligations in the natives.

We drove on through the last fringes of the Noel-Baker forests, on the curve above a sea in which floated the Northern Sporades. On our last day with Margaritis we were accompanied by his brother-in-law, a prim young school-teacher who had arrived to support his sister while in labour, and who now went jaunting with foreigners. The relatives who come along for the ride, and the peasants you pick up for a lift, are part of the entertainment when you travel by taxi in the remoter Greek provinces.

We swirled down finally into Loutra Aedipsos, routing memories of the past. Certainly the stench of sulphur remained, but the fleshly present had taken over the shady promenades and grand hotels at the height of a steamy summer season. Hotel receptionists laughed at us. The proprietors of humbler rented rooms with sand-heavy bathers and towels hung to dry on balconies, smiled us out of their narrow halls into the street. On the outskirts of the town we were offered a couple of stretchers in the tower of a rust-coloured building designed round a lift-well earlier this century. The stains on the bedspreads of our meagre lodging and the sound of the lift rising and falling in its iron cage decided us to return to Athens. Margaritis offered to drive us there, but we sent him back to the nursing mother, and waited for the ferry to the mainland. Only then it occurred to me, waiting to cross the narrow strip of water, that Euboea is really an island. I might have started writing the story *Fête Galante* already kicking inside my head, but we had no lodging, only the bus-cum-ferry station. So for once I had to contain myself and one of those island stories till our return to Athens.

Before our visit in 1975 the Polymeropouloi had moved house. They were no longer in possession of the small rooftop flat where Manoly and I had led our post-War Athenian life. Elly had bought another flat, again on the roof, in a remote workers' suburb, and this had become our substitute home. In spite of the business of getting there, we were always glad to return to it and close our door. From one side we looked out on a scarred flank of Hymettus, on the other to Aiyina and the Saronic Gulf whenever visible through the Athens smog, but we were still able to see, above the roofs, amongst the fumes, what looked like a fragile toy made of matchsticks – the Parthenon of blessed memory, where I used to walk in undisturbed solitude at the end of the German Occupation and the Civil War.

At times the noise of present-day living almost blasted us out of the rooftop flat in the Street of the Sirens. In the skeleton of the apartment block going up beside us there was always the yammer and vibration of electric drills and jack hammers, and the voice of a young workman singing about the widow who invited him in to fuck her. Sometimes a terrified sparrow landed on our terrace attracted by the crumbs from our breakfast, but was quickly sent rocketing into space by a cement clod from the building site next door. The average Greek can never let bird, cat, or dog exist in peace.

The other side of the canyon dividing us from the opposite block, human fowl, restricted by iron balconies and the small concrete cells in which television constantly flickered, led nervous, livery, isolated lives, remote from those community interests they had once enjoyed in their villages. They eat better, admittedly – if chunks of martyred Argentine beef, wads of congealed macaroni, and mounds of flabby fried potato, are better than a crock of beans, tomatoes, onions, and a draught of harsh red wine on the edge of the field in the mid-day shade. Are they happy, these peasants of yesterday, now referred to as 'agrarians'? They don't seem to be, judging by the screams which lash out from some of the cages, the wives seldom visible (the same drudges in a different setting) husbands swinging turbulent *komboloyia* on their return from work in jampacked buses, and the little bandy children pressing midnight faces against the bars.

We made friends as we always do, somewhat shyly, with the Anatolian grocer, the chemist from Constantinople, and the fat, steamy dry-cleaner from a village in the Western Peloponnese. We washed our underpants and socks, and hung them out to dry in full view of that matchstick model of the Parthenon. We breathed an air laced with cement-dust and petrol fumes, and felt ill, Manoly from his arthritic pains, I from a chest I could not bring myself to confess to. We got drunk at appropriate hours on the bottle of ouzo kept in a cupboard in the kitchenette, and I wrote the story *Fête Galante*.

The Mainlands

Greece is the greatest love-hate for anybody genuinely hooked. We who are, amaze amateurs by giving them glimpses of what we see and feel. In the same way foreigners who have spent five minutes in Australia and grown sentimental over their superficial acquaintanceship think you have gone off your rocker or that you are another of those Australians with the so-called inferiority complex if you tell them the truth about it. What you truly feel about a country or an individual of great personal importance to you, generally shocks when you are honest about those feelings. If you are pure, innocent, or noble – qualities I don't lay claim to – perhaps you never develop passionate antipathies. But Greece is one long despairing rage in those who understand her, worse for Manoly because she is his, as Australia is worse for me because of my responsibility. There are times when I think that M. in his willingness to forgive Australia must be more Australian than I. But he only came here and joined a club, as those of the first generation inevitably do. He and I won't breed another generation unless those who read and understand my books. I believe that books *could* breed future generations in spite of the pressures on Australian children to choose illiteracy and mindlessness, or if home-bred totalitarians and foreign invaders do not destroy our tentative Australian literature.

Greece is mindless enough, unless when it comes to politics, and there confusion abounds. The Greeks are all for chatter, and political chatter is the most seductive means of wasting time. Few Greeks read anything but newspapers, though literary coteries abound, usually revolving round the poets. If you ask why there aren't more novelists you are told the political situation makes it dangerous for Greeks to write novels. As so many Greeks have been prepared to die for their political beliefs, I can only feel they are not prepared to forgo their chatter for the long and arduous stretches of seclusion which the writing of a novel demands. The chatter is amusing, stimulating, till you begin to see it as another kind of drug abuse. Then you long to escape from its clutches.

The mainlands with their bus networks, their rutted roads, the long pauses when you wait for something to happen, the diarrhoea, the leap through a gap in a pine forest into an expanse of blue, a posse of neo-Fascist guards going through our papers at a frontier post, a steely lake which has received the bodies of Greek women sewn in sacks by the occupying Turk, memorials in villages where almost all the inhabitants were exterminated by German overlords, a grey amphitheatre deserted except for a giant Roman snail on one of the seats, till a tattered poster obtrudes and Christopher Plummer's Oedipus rears his over-lifesize head with that possessive arrogance of the commercial West — incidents and images such as these have helped temper my passion for Greece; they have forged a relationship without which my life would have been sterile indeed.

Manoly remembered Pelion from his boyhood. He had made it sound a promised land: a mountain rising out of the sea, furnished with deciduous forests, and towards the summit spruce and pine. In those days you reached the sea or climbed to the heights on muleback. When we visited Pelion for the first time together in the spring of 1963 we contracted for a pair of mules to take us one morning to the sea and back, but otherwise we were whirled around the mountain over give-and-take roads by the inevitable loosely articulated taxi. The nights were still cold at Tsangarada. At the village *kapheneion* we sat outside in our woollen wind-cheaters. Between us and the illuminated interior was a square of concrete on which dancers must have performed at another season. Opposite us, against the façade, and inside the large timbered room from which the establishment's business was conducted, the inhabitants had gathered to observe – a withdrawn, alpine race, milky-skinned, in some cases even rosy-complexioned. On both sides of the concrete square we sat sipping our *tsipooro** and nibbling *tsitsiravlo** as *mezé*. The mountain people were shy, as we were, so it suited everybody. Apart from our taxi-driver and a waiter at the tourist hotel who told us that what we were eating was 'cow with mushrooms' we got to know nobody in a conventional sense on Pelion. It really didn't matter, because silences and faces can often convey more than words.

It seemed we had the mountain, the secret villages, the peasant churches, to ourselves. (The keys of the churches were as usual difficult to contact.) The fortress-houses built against the Turk had been horribly scarred, and in some cases burnt out, in the last stages of the Civil War. Wherever we went we observed unrevealing faces observing us. We encountered dignity and often kindness. Those of the Greek peasantry who have resisted selling their houses and land, and who have remained locked into their traditions on island and mountain fortresses, are the true nobility of Greece.

Some years later, on our second visit to Pelion, it was another season, a different climate in more ways than one. It was harvest-time for apples and chestnuts. Whereas in spring the inhabitants seemed barely to have thawed out after winter and were inclined to look fearfully over a shoulder as though expecting a cold footfall to creep up on them, now as they stood loading their trucks beside the suicidal roads they appeared to have taken on some of the character, the texture, the perfume of the great crimson Pelion apples, though at the summit, as we ate our sausages and bean stew, and drank draughts of icy mountain water, there were still the silent crop-headed, milky-skinned alpine types. Otherwise around the mountain an invisible harvest dance was in progress, in a perfume of apples, and stench of nutty armpits. The key to the church was more than ever difficult to come by.

This time we stayed at Portaria in the usual stereotype Greek tourist hotel. Everywhere we were only just tolerated because everybody was suspicious. We used to walk to the village of Makrinitsa hoping for a glimpse of the Theophilos murals we knew to be there. In Makrinitsa, supporters of the Junta had come out in force; banners strung across the village street hailed the Colonels with true Fascist fervour. No one was prepared to show us the Theophilos murals. Some did not know of their existence; others told us the keeper of the key was away in Volos.

We made another attempt to see Theophilos murals in a village at the foot of the mountain. They were in a bakery, where the baker's wife shrieked at us that she wouldn't let us in unless we guaranteed to buy them. The baking of bread is a serious business in Greek villages. Even so, the woman's extravagant proposition seemed to me that of an unreasonable shrew, till in recent years I saw in Athens Byzantine murals peeled from the walls of mouldering provincial churches and those taken from an archaeological site in Santorini. The baker's wife, a practical woman, was only right in shooing from her doorstep the pestiferous tourists who interfered with a baker's working life, just as I consider it my right to dismiss the journalists and thesis-writers who would wreck those other loaves I am kneading and proving in my work-room.

Since our two memorable visits we have not been back to Pelion, but hear that blocks of flats have risen at Makrinitsa, and on a night in 1979 at a cinema in Sydney, we were introduced to the horrors of a Pelion ski resort in a film to promote tourism in Greece: chair lift, fun on the run, snow bunnies, arrogant randy males, monied adolescents taking everything for granted. I tell myself I must not hate human beings. I try to conjure up my vision of an actual landscape and the inhabitants to whom it belongs. But it is hard for visions to survive in the plastic present, as mascara trickles from smeared eyes and blown-up lips gorge themselves on mass-produced food. There comes a moment when a stream of semi-digested eggplant, mincemeat, and tomato is vomited across the screen of memory in a sour splurge.

We first passed through Metsovo in the days before it was threatened by progress, or Western decadence, whichever way you look at it. Like Pelion, the Pindus has become a ski resort. Fortunately Metsovo itself, built on a ledge between a mountain and a precipice, may escape physical if not moral destruction. The inhabitants are more reserved than most Greek communities, perhaps through the Vlach influence. A Vlach is a phenomenon nobody can satisfactorily explain, a nomadic race of shepherds and herdsmen wandering from Asia Minor through Northern into Middle Greece, their language closer to Romanian and Latin than to Greek. Today the inhabitants of Metsovo are settled, but Manoly could not understand their dialect. On one of our visits we overheard a conversation between a middle-class Greek tourist from the south and a peasant of the Pindus. The Greek was confiding in the local that he had something of the Vlach in him, while it was obvious the peasant did not consider the foreigner a candidate. The Vlachs, we could see, had preserved one of the closest-knit secret societies. It is easier to win the confidence of a Freemason, a homosexual, a Jew, seldom a Frenchman – but probably never a Vlach.

It did not worry us. We loved Metsovo. It had a pastoral air. Handsome women dressed unselfconsciously in their local costumes strode through the evening arm-in-arm with their more sombre men. One remembers grassy slopes of an alpine green, a smell of burning charcoal and grilling meats. We found an inn run by a couple from Andros, as far from the Pindus as you could get. They took pride in their table. When we asked what we could have to eat, they told us, '*O ti thélete.*' (Whatever you want.) Normally this statement is followed by rejection of everything you ask for: it is not the season, it is finished, and so on into the catalogue of excuses, punctuated with clicks of the tongue, that final expression of the Greek negative, and the Adam's apple performing the functions of a rapid lift. But in this instance, miraculously, there was indeed every dish we had ever dreamt of. We feasted, as did any other Sunday traveller through the Pindus, in the long hall opening at one end into the gleaming pan-filled kitchen containing *o ti thélete*, at the other, lush sward under summer trees.

On our next visit to Metsovo, as we approached from Thessaly neither on foot nor muleback as favoured by dedicated travellers, but making our own contribution to Western progressive decadence in a Metaioran taxi, there were still hummocks of dirty snow cluttering the forests and filling ditches alongside the road across the Pindus. Metsovo was not at first visibly changed, though the inhabitants had not yet fully shaken off their enforced hibernation, which they spend weaving, carving, and getting children in houses buried under the snow. As we panted up and down the steep stone streets, the usual anxiety gathering in my lungs, we were observed with a suspicion which could have been of Vlach origin, or else imagined. At our alpine hunting lodge or inn, for which we had made a beeline, our doubts became increasingly serious and palpable as we crunched over grit in the room where we had slept the time before, inspected sheets which showed signs of other recent travellers, and the plumbing (never brilliant) which by now had practically given up – like our Andriot hosts. This well-fleshed, hospitable couple had melted to shadows of their former selves. The woman seemed embarrassed when I referred to the feasts we had enjoyed at her well-stocked table. The man, grown surly, avoided us. At night in the dimly lit, practically deserted hall, emptiness and a concrete floor acerbated voices and amplified the grating of chairs, the clatter of plates and cutlery as we were served the most abject meal. We never succeeded in accounting for the Andriot couple's decline, whether through menopause, the break-up of a marriage, hostility of a mountain community to islanders conducting a business in their midst. We did catch a glimpse of hostility towards her employer in the local girl whose duty was to clean our room, which she kept scrupulously dirty, whether because her mistress was a shrew or a foreigner we could not tell. But this time in Metsovo there was an overall sullenness, contempt, cynicism, difficult to analyse. Was it the cynicism of political despair? At the bottom of the main street where it widens into

grassy parkland there were posters proclaiming the Colonels'
Revolution. There was no sign that the splendid women and
their sombre men taking their evening stroll as usual accepted
or rejected the situation. Veils of Vlach dialect hung between us
and any possible estimation of the truth. But there was a pall on
Metsovo.

We both cherish a memory of what will have been our last
visit: that of the bakery at night, an illuminated cube in which
sat a knot of aged, black-kerchiefed women knitting and gossiping, for the baker's shop in a Greek village is the women's club.
Through the open doorway poured the smell of baking bread,
which will always convey, to me at least, one of the few
convincing messages of goodness, hope, and possible continuity
in a threatened world.

The Northern Reaches

From a terrace in the northern town of Verroia you can look
down on the blanket of purple smog which now hides the more
familiar south. We felt cut off, estranged from civilisation, a prey
to the Slav overtones which weave through the life of northern
Greece. Manoly has nightmares in which he is chased by Turks;
in my case a Slav seepage rises out of the depths of the unconscious. These double and differing anxieties led to considerable
disagreement during our travels through the north.

We had gone to Verroia chiefly to investigate the coven of
Byzantine churches hidden from the Turk in the alleys of that
provincial town now proclaiming itself a city on the strength of a
handful of jerry-built apartment blocks. The churches were
elusive as always: too dilapidated to be shown, or the Orthodox
authorities declined to give up the keys to infidels, heretics, or
possible vandals. We sat on the terrace above the canopy of
southern smog, sipping our ouzo, surrounded by the usual
provincial ladies dreaming of next season's handbag and the
officer they still had to meet.

Our big aesthetic experience in Verroia was the display of

vegetables — semi-precious eggplant, capsicums, onions, to-matoes, okra, thrown in casual mounds in markets and shops by those who did not appreciate their value. The vegetables often persuade us we could take root in this or that grisly Greek provincial town. But we flung on by bus from Verroia through kilometres of apple orchard, boring in the long run, to Naoussa, where we took a room for the siesta, then on to Edessa of the waterfalls.

Edessa was full of Greek tourists and the plastic clichés of tourism. Aqueducts guide an abundance of water through the town towards an operatic climax in which it hurls itself into cataracts, like fine-spun grey hair tangling with the branches of dark trees, before disappearing into a void as inevitable as that chosen by suicidal poetesses.

We left Edessa in a taxi, the owner of which had agreed to drive us through the wilder stretches of our northern travels. Coniferous forests, cover for bears and wolves, sparkled in the green, early-morning light. Lakes glinted. The frontier with Yugo-Slavia was, so to speak, over the hill.

It was at Florina that the Slav North and the Colonels' Greece made their real impact. As we drove into a small aseptic township, the inhabitants of which get their livelihood from the manufacture of Fanta cordial, the Colonels' troops showed themselves, young crop-headed men in berets and razor-edged battle-dress trousers. Our papers were seized and carried off for investigation. As the atmosphere thickened around us I was reminded of pre-War experiences in Hitler's Germany. Checked and stamped, the papers were returned eventually, but suspicion followed us, not without reason, down the road to Kastoria.

We dived down alongside another of those steely Greek lakes, like the one at Jannina which always makes my blood curdle. Jannina, with its history of Turkish brutality before independence, is for me less ominous than the damp threats of Slav penetration which hover round the lake and over the town of Kastoria. Everywhere damp as we plodded up and down the steep streets and fossicked for the ancient fortress-houses, in this instance scarred not so much by civil war as present-day van-dalism: plaster stripped down to a corsetry of lath, and farther still, in some cases exposing the skeletons of what had once been noble forms.

A surreal touch was added by the snippets of fur lying strewn in the steep damp streets, suggesting a midnight slaughter of squirrels and cats. But actually, Kastoria is a centre of the fur trade. For our evening meal we ate putrid trout which must have spent several days in their pools of oil in the taverna. We slept uneasily in damp beds and rose before dawn to catch a bus for the south.

This was one of the more memorable occasions when we spent hours waiting for the connecting bus at a crossroads, in this event Para — cryptic name. Oh, the humbler Greeks are accustomed to wait, the peasants with their baskets, the black-kerchiefed grandmothers bound for Perth, Adelaide, Melbourne, Sydney, not forgetting Brisbane, without an inkling of what is in store; all they have is their trust and their demanding families. We have waited with them and this innocent trust has racked our souls. Nobody who has sat with Greek peasants at the roadside or in one of the ferocious world airports will be the same again. Nor anybody who has suffered paradiarrhoea in a filthy Para lavatory, waiting for the bus which doesn't arrive. It did at last, and we continued burrowing into the unending Greek landscapes, half actuality half dream, through Kozani, Larissa, towards the south.

Last night, of fitful sleep in my bed at Martin Road, I found myself at Mount Victoria, a similar mountain green, a physical asepsis not unlike that of Florina. What can't happen here had happened, in Australia. The berets, the knife-edged battle-dress trousers were stationed at a bend in the red road I have known all my life. From against a clump of black-green conifers the guards stepped out and asked for our papers. On this occasion we could not produce them . . . .

Towers, Stones, and Dust

The least accessible, and traditionally most ferocious part of Greece, the Mani, had been opened up by the time we went there. Compared with other philhellene travellers and authorities, that courageous Romantic and eternal commando Patrick Leigh Fermor, and the stern, masochistic governess, our good friend Robert Liddell, Manoly and I are decadent explorers expecting a modicum of civilisation for our pains. That we almost never get our modicum stanches a feeling of guilt I might otherwise suffer from.

We decided to approach the Mani by sea, making landfall at Monemvasia. I had not set eyes on the glittering mineral coast of the Eastern Peloponnese since leaving Greece, disconsolately, for demobilisation in England at the end of Hitler's War. Then winter, now summer, the mountain ridges had not quite the same cutting edge, the sea a less forbidding glassiness. As for myself I had achieved what I had most hoped for. I don't mean worldly success, which is equal to moral suicide unless you are determined to survive, but a lasting relationship with a human being I can respect and trust. (To be able to rely on *one* human being in 1981 is reason for belief in miracles.) To that extent I felt fulfilled, if also inclined to grizzle at finding myself embarked on another voyage in a smelly Aegean steamer. It cheered me up to talk to a man with a brace of pointers returning to the island of Kythera who turned out to know Kytherean friends and neighbours at home in Sydney. Talking about these Austral Greeks with the Kytherean huntsman was curiously comforting, as though I had arrived at a synthesis of the far and the near, Europe and Australia, opposing elements in myself. It was a situation I could not have imagined in my egotistical, ambitious youth, or my emotionally chaotic middle-age.

At Monemvasia we were unloaded from the steamer into a small restless boat which ferried us as far as a mole. Choppy waters slapped the mole and drenched us as we landed. It was all part of the performance. We were less concerned about the soaking than finding transport from the mole and a room in which to spend the night. For the moment a squalid modern village was of greater importance than a rocky fortress embodying the past. After being turned down at every possible lodging we were received into one of those concrete 'hotels' which have sprung up all over Greece with promises of modernity. In fact the Greeks are never more primitive than when aspiring to be modern. Nothing ever works. As we were setting out at first light it did not matter all that much if the hot tap ran cold and the power went off during the night. When ageing bladders drove us out on to the landing in the dark the lavatory was easily located by its smell. When we were shown our room on arrival, electric drills were still at work alongside, where yet another concrete block was going up as a tribute to modernity.

As soon as we had settled in we started the walk along the causeway to the rock fortress of ancient Monemvasia, held in turn by Greeks, Venetians, Turks, and today, it seems, occupied principally by Germans and Scandinavians, an irony as sour as the encroachment of concrete and plastic on what one would like to think a deathless country. Clambering up and down the rock, looking for the various churches, losing ourselves in the labyrinth of ruins, wheezing and spitting in my case, suffering arthritic tortures in Manoly's, it was inevitable that we should vent our feelings on each other at times. We have always done so, and always shall. It is our way of survival. He forgives me the Romanticism of which he has never altogether approved, and which is probably a symptom of spasmodic asthma, as I forgive him the rigidity of an arthritic Orthodox faith.

We returned to the modern village to forage for food, and after considerable search sat beside the sea, still dominated by the rock fortress, our plastic table-cloth twitching in a gritty wind while we waited for our humble expectations to be fulfilled, as indeed they were in the usual slightly nauseating fashion.

It was still dark the following morning when we stumbled out of our sleeping hotel into the street. We had been warned the evening before that seats on the bus could not be reserved and that we ought to be on time to avoid missing out. We were. A friend once accused us of arriving at airports with time enough to read *War and Peace* before the plane takes off. Monemvasia spurred us into living up to our rule. At the end of the street and the causeway the presiding rock was barely visible, its ancient, rather sinister mass like a form created by some Twentieth Century etcher. To put in time we looked for breakfast, and found at the top of the village street a *kapheneion* made frowsier by the early hour and proprietors just risen from their bed. Even so, we were thankful for our lukewarm Nescafé and the sesame seeds on some stale rolls. As the proprietors moved behind and around the misted glass cases containing the dead food they had for sale, we were content with our small mercies. One of the pleasures in Greek travel is to sit mopping at gummy eyes, collecting those sesame seeds on the tips of what at other times would be disdainful fingers, while dawn unfolds to background music from a hissing primus.

By the time we returned to the bus agent's office the rock at the end of the causeway was less an etching than an aquatint, the threads of a few frail reeds trembling on the foreshore, faint colour and birds on the wing invading the colourless sky, passengers gathering, sleepily croaking, coughing, hawking, lugging their plastic carrier bags, dilapidated cardboard suit-cases, and traditional osier baskets, their contents stitched down under a calico topping with name and destination indelibly marked. We had no trouble finding seats, and started on our long cross-country drive to Gythion, which we had chosen, more or less by taking a pin, as the centre of our travels through the Mani.

Gythion turned out a somewhat unprepossessing town, with echoes of the Colonels and the Monarchy, and some of the worst plumbing and food. There is a small island, Cranai, where Paris and Helen are said to have enjoyed each other after their elopement. Today the island is linked to the town by a causeway, and at this period had become the Mecca of German hippies with camper vans. It was littered with rubbish and human shit. Still, we enjoyed climbing the terraces of Gythion, asking directions and general information of friendly women, and sipping our ouzo in a cool breeze beside a sea which smelled unavoidably of sewage.

Again we came to an agreement with a taxi-driver, less ingratiating than our favourite, Margaritis of Chalkis, this one an elderly, authoritative, desiccated bloke, the Greek version of an Oz Digger, who drove us almost daily and put up with what he must have thought our peculiar wishes and objectives. There was not a village in the Mani where he did not have acquaintances; he knew every peasant to whom we gave a lift, usually shy women and little girls who barely whispered answers to his rambling inquiries. The driver was as anxious to promote

himself through the foreigners on the back seat as the gratuitous
passengers were tactfully determined to resist being drawn into
what their delicacy had decided against.

Despite a ferocious past, a baked, drought-stricken landscape, and the Germans who are taking possession of it today,
the Mani is a noble corner of Greece. Poverty and stones did not
prevent us realising afresh that the true aristocracy of Greece are
the peasantry who have stuck to their land: static, wary men,
with eyelids like those of the lizards scuttling through their
stony fields, and ravaged, tireless women. The towers from
which the Maniots defended their estates and conducted their
bloody feuds are crumbling now. Perhaps it is all to the good
that the Germans are moving in. They will be absorbed like
other invaders after they have made their contribution. For all
their passionate hatred of the foreigner, the Greeks have the
power to assimilate and come out on top. Or am I too hopeful?
Remembering the stone faces of the old men, the plodding
women, the towers fingering a sky drained white by heat, I
don't think so.

Maligned by historians, travellers, and Philathenians, Sparta
today has it all over Athens the 'city of sweetness and light', the
'violet-crowned' etc. etc. However abominable its totalitarian
past, Sparta has a peaceful air. Its best houses, dismissed by
guide books as being of no interest or importance, have the
noble proportions of Greek Nineteenth Century architecture,
almost all examples of which have been destroyed by Athenian
developers in their rampage after quick money. Not much
interested in classical Greece, I have felt each time we passed
through Sparta on our way to Mistra that I would like to return
there for a few weeks, to read and drowse and try to forget the
mess the country has become.

Only an amateur, I cannot begin to deal with Mistra, that
'cheese' or hive of Frankish and Byzantine history. The first time
we approached this magnificent mound was through a grey
spring drizzle on a conducted tour for culture addicts, on that

occasion almost all of them Americans. I am not good at absorbing culture. I cannot listen to lectures. I am barely educated, a fact that many are not prepared to believe: haven't I been to Cambridge? haven't I got a degree? didn't I win the Nobel Prize? What they cannot understand is that none of this matters creatively, and that the monotonous surge of information at Mistra or anywhere else causes me to draw apart for the same reason.

Manoly and I are at our edgiest on Mistra, what with my inadequacies as historian and theologian, and the Lascaris connections, the crumbled palace, the ancestor in the chapel of the Pantanassa, his eyes gouged out by the Turk. So we lose each other on Mistra for fear that our edginess might drive one of us over the edge.

On that first occasion we dripped around in our raincoats, apart, and came together again with our fellow tourists in the convent visitors' room where the nuns got us drunk on ouzo and then sold us embroideries. We still have the little mat in the icon corner in the house in Sydney where we continue living in 1981.

The second occasion we staggered round Mistra was years later in the blaze of summer – the lizard season. We were brought there by our taxi-driver, the elderly Gythion version of the Oz Digger. We began at the top of the mound, separated, lost and found each other among the ruins, sidled past the sign now pointing to the site of the LASCARIS HOUSE, visited the Pantanassa where the nuns had become inhospitable and sharp; they complained that the tourists were using their tap and leaving the water running. But their convent was spotless as ever; the world of plastic had not yet corrupted them.

Looking out from the historic and architectural muddle of this great mound of tumbled stones and drowsy lizards I experienced one of the brief moments of perfection Greece offers the obsessed. Our sweat, our aching feet, the plain of Sparta stretched out in front of us like a dusty carpet under a heavy heat haze, Taygetos puritannical without the snows and violet light of our first visit, were all part of the experience. For the rare glimpse of nobility, the grudging kiss, all you have suffered in the way of caprice, trashiness, a degeneracy of the beloved, is at

21 Fran the third niece and great-niece Catherine Bishop.

22 A writer's life as it never is, Martin Road 1973.

23 Matt and Flo Davies in old age, Mount Wilson.
24 Trudging round Toledo, 1975.

25 Protesting against the Olympic Stadium which would have wrecked Centennial Park.

once acceptable; your passion is justified.

After lunch we drove down into the village and out the other side where a *paniyiri* was warming up. I cannot remember which saint the fair was celebrating, but there were horses, gipsies, 'Christians', the jingle of harness, the babble of voices, music from a roundabout, and the first wavelets from a plastic sea. Flowing around and over us this fairing was another instance of the Greek dichotomy of earth and spirit.

When we left Gythion for good our driver took us by the coast road to Kalamata. I suspected it was to bring me together with Leigh Fermor, whom he professed to know, along with everyone else in the Mani. On the way we stopped at a village where he introduced a young woman who had lived in Australia but returned to Greece for some family reason. In the village where she had been born she was obviously wrung by nostalgia for the hairdresser and other Australian 'amenities'. Our attempts to reconcile her to the distinction of her village naturally failed.

We continued to Kardamyli and the confrontation with 'Paddy' which our driver was determined should take place. I dread meetings with those I admire, and as we approached the house was relieved when the driver told us, 'Paddy can't be here. His car isn't.' Leigh Fermor was often away from home since starting treatment for throat cancer. The village where he has built his house after an arduous life as soldier, explorer, and writer, has a gentle, pretty charm compared with the scorching Maniot villages to the south. Kardamyli suggests Italy rather than Greece, at any rate the diluted Italianate of Corfu. Stretched on its promontory, the distinctive but atypical house looked idyllic as we swept down towards it. Lucky the man who can choose an idyll for his old age. I wish I could, but my life continues to be chosen for me.

Kalamata of the olives and the oil is a sympathetic town of no great interest. We enjoyed poking about, looking again at shops however mediocre, and finding a restaurant where, in the dusk, we sat at a table on a square, in the shadow of a mountain, and ate food of surprising excellence after the filth at Gythion. Nor did we see a single foreign tourist. There aren't enough ruins in Kalamata.

Exhausted to the point of longing for Athens, the Polymeropoulos hospitality, and our rooftop flat in the Street of the Sirens, we would never have seen Pylos and Navarino if our often irritating taxi-driver had not stayed on for reasons of his own and insisted on dragging us for a last expedition the following day.

The painter Brett Whiteley asked us recently which we considered the wonders of the world. It is the kind of question which always strikes me dumb. I could only mumble Ayia Sophia and the Parthenon, without being able to add to them. Till Manoly suddenly united with me in an outburst of, 'Pylos! Pylos! Oh yes, Pylos and Navarino Bay!'

At any rate on the evening when we saw it the landscape was the most sublime imaginable, the landlocked bay almost a circular lake. If it had been a lake, the waters of Navarino would have taken on that dead, steely look which seems to distinguish Greek lakes of any size. Instead, as a bay, and in spite of an oil refinery sprouting like a cancerous growth from the inner shore, it has remained alive, a mirror flashing every letter in the code of beauty. The great battle fought in the bay, and which contributed more than anything, if also by accident, to Greece's independence from the Turk, may have added lustre to its eye and my romantic vision of it.

If we did not visit the palace of Nestor (Robert Liddell's 'Homeric bore') it was because I am less interested in ancient Greece, property of the fossickers, the professors, and the tourists, than in the Byzantine and Frankish versions of it. We did go as far as Methoni to the south, its magnificent open fortification a stage awaiting some spectacular theatrical event.

On the drive back to Kalamata from Navarino, and Pylos the white, dreamt-of village, we passed peasants returning from the fields leading white house-goats staggering along under bursting udders. The whole landscape was heavy with fulfilment. We too, felt fulfilled, as one does in Greece at moments when earth and spirit complement each other.

Next morning we caught the bus to Athens on a stretch of

barren ground beside a dry river, slightly different from the typical Greek bus station, but every bit as depressing. There were booths selling mummified, fly-haunted food, multi-coloured scarves and handkerchiefs, and the usual plastic trash.

Then as the bus got under way and mounted the road, a splendid, fertile landscape surged around us on either side. We settled down till at Megalopolis a cement works out of science fiction spewed clouds of dust and fumes over the plains, olive trees, the mountains of the Peloponnese. From Megalopolis to Tripolis I had beside me a small fat grey-haired man in black suit and dark glasses reminiscent of the 'great' Austral Greeks. We somehow resented being thrown against each other by the movements of the athletic bus. Our bodies told us we were on opposite sides.

At Tripolis, the prototype of Greek provincial towns, we shed the pseudo-Austral Greek. We ate passable *tyropitta*,* and forged on into a distance which drained us of all resistance, and finally life. Our mere bones were jogged, tossed towards the roof, and returned to seat level for further torment. There was a staging post somewhere along what used to be the Saronic Gulf where we were disgorged, and allowed the time to feed if we had the desire. Any appetite left us after a glance at the prospects. While we were standing on a boardwalk above the sadly humiliated waters of that classic inlet, I glanced down, and there amongst the rubbish was a plastic spoon stamped with the word AMERICA.

We reached Athens, and a day or two after our arrival, read of a brilliant government plan to 'develop' Navarino by building shipyards and extending the existing oil refineries. As a *coup de grâce* the planners envisaged a site for the Olympic Games at Pylos. Mercifully the project sank, but during the world crisis of 1980, Greece has again come up with a plan for a permanent site for those renowned trouble-makers, this time at Olympia itself. Rather the death of the Games than the death of Greece. Not only on a plastic spoon casually dropped beside the Saronic Gulf, AMERICA is writ large across its victim. It is tattooed into the body of a goddess turned prostitute, by poverty, materialism, and international politics.

Episodes
and
Epitaphs

1 *Luncheon on the Yacht*

In 1963, during Queen Elizabeth II's visit to Sydney, I received
an invitation to luncheon on board 'Britannia'. Surprised and
flattered, I prepared to satisfy my curiosity. My political beliefs,
while forming vaguely, had not yet firmed. A few years later,
conviction and events would not have allowed me to accept.

At that time it was incredible to be setting out from the dust
and heat of our failed farm at Castle Hill, in a hastily bought
best suit, to lunch with the Queen. I was very nervous approaching the wharf where the yacht lay moored, though my courage
returned, and with it an objective cynicism, as I realised I was
far less nervous than some of my fellow guests: a newspaper
magnate and his third wife, a refrigerator manufacturer with, I
am sure, his original, a swimmer, a noble lady related to the
Queen and married into the colonial gentry, an admiral and his
blue-stocking, with various other odds and sods. I almost forgot
the most important guest, Joern Utzon, a tall man as handsome
as they come, but handicapped by the English he spoke, which
was as boneless, and to me as unintelligible as his native Danish.
At least I enjoyed trying to talk to Utzon, and I think we both
enjoyed standing together uneasily at one end of the saloon. At a
later date he showed Manoly and me over the foundations of his
great opera house while it still looked like the ruins of Mycenae.

As I set foot on deck an amiable Anglicised equerry assured
me he had read *Voss* and that a copy was lying on Her Majesty's
bedside commode. I realised quite soon that I had come wrongly
dressed, but increasing assurance and indifference helped me

resign myself to my mistake. Having to buy a suit for an important occasion after living mostly in boilersuits amongst dogs, goats, and vegetables, I did the wrong thing. I must have looked like an Italian waiter on his day off, far too tight, and far too black. Or was it *Voss*, lying on Her Majesty's bedside commode? Anyway the Duke, dressed in the tweediest hacking jacket, gave me his best Glücksburg scowl and made a beeline for the swimmer. He appeared fascinated by this swimmer, who was, it turned out, a very civilised young man.

Before lunch we were dragooned into confronting lines in preparation for inspection by the Monarch. The newspaper magnate's third wife who was starting her remarkable climb up the social ladder looked the colour of bottled peas. I felt genuinely sorry for her; I might have looked as green myself. Eventually Her Majesty appeared, and strolled between the ranks of her erect colonial subjects distributing gracious if somewhat glacial smiles. As she passed, the ladies attempted a curtsey. The refrigerator-wife almost keeled over and had to be pulled back into balance. Miss Doris Fitton of the Independent Theatre gave a splendid rep performance.

At lunch I was seated between the noble lady related to the Queen, from whom it was almost impossible to extract a word, in spite of, or perhaps because of, knowing her husband's mother and sister, and on the other side, the Duke's swimmer.

Murray Rose achieved publicity and fame early on. His mother had reared him on seaweed, we were told by our kitchen radio. It was perfectly true, and led to some dietary complications during the royal luncheon. After finding Murray Rose a sympathetic companion on what could have been a dismal occasion, I would have liked to follow his career, but he seemed to get whisked away to California by the seaweed-advocating mother, and not much has been heard of him since.

After lunch we stood about, relieved, sustained by drink, and taking some more on board. The Monarch moved among us, while the Duke buttonholed the swimmer. After a while the equerry asked me whether I would like to speak to Her Majesty. It occurred to me that if she wanted to speak she would ask for me, but perhaps that unreadable *Voss* on the bedside commode was holding her back. When brandy decided me: what the hell, here I am in this tight black dreadful suit – why not?

So the equerry led me up to join an established knot which included the admiral and his blue-stocking wife, an economist what's more, who obviously resented an outsider's intrusion. They were going on and on about the yacht's stabilisers, and how on a previous visit the Admiral had shown Her Majesty the Barrier Reef. There was no sign that she was ever going to address the writer brooding at her elbow, so again I thought what the hell, and decided to break protocol. I told her that some day she ought to have a look at Fraser Island, and asked her whether she knew about Eliza Fraser and the wreck of 'Stirling Castle'. It brought at least an appearance of life into the proceedings. 'Oh, yes,' she piped in her high-pitched, cold, china voice, 'The Naked *Lady*! We saw some of her – in *Adelaide*.'

The occasion was practically at an end. The Monarch and her consort took leave of their guests. The couple turned. They made a slow exit up a flight of stairs, she the frailer for some arrangement of hook-and-eye which emphasised the vertebrae under her pale-blue-specially-for-the-Colonies dress, he more than ever a Glücksburg bully apeing the English in his tweedy hacking jacket.

I had done it at least. After the usual interminable journey by train, bus, and on foot down Showground Road, I arrived home. It was a relief to get out of my tight waistband and sweaty shirt, and cook dinner for Manoly who hadn't been invited, but who was so much more distinguished than most of those who were there.

2 Incident in Martin Road and the Shocking Career which Developed out of It

This quiet island of houses and gardens between Centennial and Moore Parks where Manoly and I have lived ever since I decided to leave Castle Hill and re-settle in Sydney suffered what amounted to an earthquake shock during 1972 when a politician keen to advance his career and an equally ambitious Lord

Mayor tried floating a plan to build an Olympic stadium between the two parks. It would have involved bulldozing all the houses in Martin and Robertson Roads. The stadium would have overlapped the parks on either side. Centennial, with its proliferating bird and plant life, was threatened with invasion by freeways and an Olympic pool at the wilder end. Moore would have been given over to lesser athletic fields, and parts of the humbler suburbs beyond demolished to make way for a village to house athletes. We were suddenly made personally conscious of the curse the Olympics have become in the over-populated cities of today.

We the bourgeoisie on the secluded island between the parks were bludgeoned out of our complacency. It was even more distressing for those living in the suburbs to the west and south, not only threatened with losing their homes, but with the destruction of parks which had been breathing spaces for themselves and their children. After the news hit us, everyone seemed to be rushing in different directions, babbling, attending meetings, signing petitions. Fortunately those competent to organise an opposition were busy doing so behind the general confusion. I was not one of them. Local government had always been a mystery I could not have begun to solve. Now I was a dead loss. There seemed nothing I could do beyond contribute some money.

Then one cold windy day, the kind that brings despair at the best of times, I was stopped in the road in front of the house by our neighbour Theodore Simos and an academic whose name I forget, who was laid low by a heart attack in the early stages of the campaign. Theodore is a barrister, correct, conservative, born in Australia of Greek parents, which has always kept me from blowing up when we disagree most intensely.

On this uncharitable afternoon we stood discussing the two rallies organised, one in Centennial Park itself, the other at the Town Hall, to protest against the building of the stadium. Various people, political and conservationist, had agreed to speak. Suddenly Theodore announced, 'You should speak, Patrick. You'd carry a lot of weight.' It seemed to me an incredible suggestion, particularly coming from a professional lawyer. I think I said as much. Though I had wanted to be an actor in my youth, years of solitary introversion at my desk had turned me off. I was now horrified at the prospect, till looking at my neighbour's mouth I realised he was more terrified than I. This professional in the theatre of the Law was an amateur in the theatre of life, which was where I had it over him – though I remained incredulous; I doubted whether I could play the role being thrust on me.

The next couple of weeks were pretty agonising as I doodled and rehearsed my two speeches, one for the park, the other for the city, continually imagining everything that might go wrong. There was almost never a day when I did not hope serious illness would strike me down. I cursed my neighbour for ringing me in as his understudy.

Then the ordeal was upon us. Spring had taken over from winter as though in celebration of the event. Wrongly dressed in a thick tweed suit (I am almost always wrongly dressed) I found myself standing on a truck in this no longer familiar park, steam rising from every blade of grass in the vicinity, a blinding glare from freshly laundered foliage hiding the lakes beyond. Other speakers had preceded me, during which time I had been able to moisten my vocal chords behind people's backs from a little flask I had in my pocket. Even so, as I stood on the truck, faced with clumps of giant mushrooms, or worse, the few individuals I could identify, a soggy sheet of foolscap trembling in my hand, that reptile the mike daring me from close quarters, my vocal chords still felt horribly dry. The audience was waiting mercilessly. I began. Whatever came out of my mouth seemed to have nothing to do with me. Since the day of my initiation I have reasoned that public speaking is much like writing: some other person is responsible for half of what comes out. The difference is that in writing, the act is kept discreetly private; in making a speech, or in acting a part on stage as opposed to desk, your folly is indecently exposed.

I had finished. There was actual applause, probably for the amateur, from an audience who shared his sentiments. I had kept it short, not only from choice, but because of the more alarming prospect of the rally in the Town Hall. Those of us bound for the second event straggled towards the park gates under sagging banners. Some of us climbed on buses, though the more dedicated chose the forty-minute walk, many of them with children in strollers.

We were photographed on arrival. I still have this photograph of a fat old codger in beret and tweed suit surrounded by youthful supporters, advancing down George Street towards the Town Hall under one of our banners – a laughable picture perhaps, but one I cherish for its sheer unlikelihood.

Once inside the hall, and in spite of the fact that I had been going there all my life, the portentousness of its vast brown expanse cowed the spirit. So I prowled up and down the corridors, in and out the lavatory, and took a few more swigs from my little flask. I feared my speech would not match this civic solemnity, but it was too late to mend matters; what I have thought out and written down always tends to set like sculpture. The other speakers started arriving, a more considerable lot for the city, including Neville Wran the future Premier of New South Wales, a high-school girl champion of debate (not a nerve in her body, it seemed), Kylie Tennant, friend and novelist, who made a speech about a peasant with his donkey which I have been trying to understand ever since, and Jack Mundey, at that time Secretary of the Builders' Labourers, a Communist who has never been accepted for the man he is (unless in this *Boule-de-Suif* situation of the stadium at Centennial Park), dedicated conservationist, idealist alas, and for me one of the most impressive Australians.

It was a long and humid session that evening in the Town Hall. A clutch of mayors from the various municipalities which would lose or gain by the stadium had to have their say. Kylie made her speech about the peasant and his donkey, the elegant Nev showed up as the professional politician he is, and Jack as a sincere man. At least for the second speech I had a lectern to still my trembling paper. The whole scene became both nostalgic and ominous as I recalled other occasions when I had come to this same ugly hall. The first time, as I remembered, was to a fancy dress ball when I had been part of an *Alice in Wonderland* set. I went as the Mad Hatter, in a hired costume, with props provided by our cook – the cup with a bite out of it, and an easier matter, the bitten slice of bread. What stuck in my memory most vividly was the image of a young woman who had come to the same ball in a pair of high-heeled shoes and a frill made from sheets of newspaper starting just above her nipples and ending at mid-thigh, the costume gathered round her waist by a narrow black velvet ribbon. She represented The Naked 'Truth'. I was fascinated by her, the high heels, her nakedness, her courage. And here we still were, years after, the Mad Hatters among us, fascinated by the flimsy, not quite assessable Truth. I tried to tell my audience some of this because suddenly it seemed I had been around in this city longer than most people. I had tramped along its hot brown village streets and watched its neo-brutal towers piled skywards in the name of progress, but never civilisation.

I don't know how much my speech did to my audience, because you can never truly tell by applause, but it did something to myself: the desire to speak was added to the impulse to write. The frustrated actor in me finding an outlet at the end? Perhaps.

But I was relieved when this civic occasion dwindled to its last gasps. I surprised a pair of architects (some of those responsible for the neo-brutal towers) by dancing on the Town Hall steps singing, 'It's ov-er! It's ov-er!' no doubt inspired by the dregs of my flask, before friends drove me away – *home* . . .

The battle for the parks was won, not through my frivolous theatrical diversions, but by the concerted efforts of local groups from all classes, and by individuals such as that dogged crusader Neil Runcie of the University of New South Wales, and Jack Mundey who, with his builders' labourers, banned the plan to build the stadium.

At once the lotus island between the parks sank back into its habitual bourgeois complacency. My neighbour Theodore Simos congratulated me on the part I had played. I shrugged off his praise with an imitation of modesty. I got on with my work. What Theodore didn't yet realise (nor did I, for that matter, or not entirely) was that he had injected me with a virus which would threaten his principle of conservative status quo.

Though Manoly and I had started voting Labour while Menzies still had us in his grip, our convictions had been comparatively passive. As time went on we began to seethe for the injustices and dishonesties of the Establishment. We accepted the lead Gough Whitlam offered us.

I did not come out openly till the Labour Rally at the Opera House in 1974. There were those who said, 'Of course he has been got at.' What had got at me was the decline in principles, still nothing to compare with the corruption in the community today. If I have moved always farther to the left, till I wish to be known as a Socialist and a Republican, it is because of this steady increase in corruption. By 1981 some of my close, once respected friends can only count as standard two-faced Australians.

Or am I a simpleton? Is it perhaps impossible to come to terms with contemporary life without developing the techniques of hypocrisy? Anyway, at the beginning of the Whitlam era we did seem to be advancing into a brave new world. It was a situation particularly inspiring for artists, till then a downtrodden minority in Australian philistine society, so it was not surprising that many of us stood up in support of Whitlam that afternoon in Utzon's botched masterpiece. It was only surprising to myself that here I was facing a multitudinous audience, as well as those we were told could not get in, who were listening to a relay on the steps and in the court. All my old fears revived, till my voice extricated itself from its source, and somebody I could not recognise took control. Then once more it was over. Faces I knew were settling in around me – Manning Clark, Neville Wran, Kate Fitzpatrick, James and Freda McClelland, David Malouf, David Williamson, and of course Gough and Margaret Whitlam, more than ever Henry Moore's 'King and Queen'.

As I drove away with Freda McClelland and Judith Wright, Judith warned me, 'Once you put your foot on the flypaper you'll never shake it off.' How right she was. It is a warning for all artists. Yet how is it possible for any but a superficial artist to live and work inside a vacuum?

So I continued coming out on occasions which seemed to demand it, even more so after the members of the Labour Government brought to power had started hacking into one another. Australians of all classes, levels of education, of the best intentions and integrity, are a prey to their native innocence. Even a man of Whitlam's intellect, wit, and capacity for leadership was brought down by precisely this strain of Australian innocence, to a level where he has remained – a wasted talent.

After the coup of 1975 and the disastrous election which followed, I have remained a Labour supporter because, however idiotically those who lead the party behave at times, and however the unions may grasp at increased material benefits, the cynical example of the ruling class in this philistine non-culture, of money, wheels, and swimming pools, does not encourage me to go along with it. I still believe in the virtues of workers as I remember them in my childhood, and as I still come across in many an individual worker, seldom in those we are supposed to respect for the glossy façade disguising a genius for dishonesty.

3 Sir and Lady

I first met John Kerr during his brief spell as chairman of a theatre board in the Sydney suburbs. The actor Alex Archdale had helped me out by playing a part in one of my plays after several others had turned it down. Though we disagreed on many points, I felt I had to accept when he asked me to become a board member of the theatre he hoped to found. I knew it would not be my kind of theatre. I knew I must appear incongruous on a board of any kind. So it was really a waste of time for all concerned, except that I met two who were to become my friends, Ruth Cracknell the actress and her husband Eric Phillips the picture framer. Then there was my encounter with Kerr, a respected judge, who later divided the country, and ended up one of the villains of Australian history, if a somewhat comic one.

When I first met him he was already a florid figure whose suits had difficulty containing him, his hair white and fluffy like a wig in one of those amateur productions of *The Rivals*. Considering his rise from humble origins (his father was a boiler-maker) to a position of some eminence, one overlooked the comic side; the snob in me even accepted the voice. But as time passed and history unfolded, I was made more conscious of the voice: it had a tinny, common edge.

It was not long before Kerr resigned from the board. Legal duties in Canberra and as Chief Justice of New South Wales took up all his time. I forgot about him. We had nothing in common beyond our tenuous attachment to Alex Archdale's Community Theatre, from which I withdrew soon after Kerr.

A fanfare of trumpets, a rattle of drums, and I realised Kerr was again on the scene: he was being installed as Governor-General of Australia. That was that. I never imagined our paths would cross (as they did only briefly and unimportantly.) At this stage there was nothing I could hold against him. I thought, better an Australian as Governor-General than another English import. We read about the serious illness of his wife, an exceptional woman according to all those who knew her. At the time I was less interested in Kerr as a symbol of office than as a man living through a tragic episode in his personal life. The wife died. The husband had our sympathy, both personal and formal.

Then, suddenly, a spanner was thrown into the formal works. Kerr, or Sir John as we must remember to call him, was re-marrying. Sir John's second Lady, not so brand-new it turned out, was the wife of a fellow judge hastily divorced for vice-regal convenience. Along with many others, Manoly and I exchanged a few knowing nods and smiles. The event did not affect us to any great extent because we didn't imagine we should make an entrance on the vice-regal scene even as spear-carriers.

Kerr had been the friend of Gough Whitlam, the Prime Minister we had supported. He was the friend of our friends James (Senator) McClelland and his wife Freda. Even so, it surprised us when we were invited to dinner at the McClellands' to meet the Governor-General and his wife.

Not long before this occasion, Whitlam and his Government had decided to introduce a system of Australian honours to supplement the Queen's British rewards, and eventually, one hoped, replace them. I had been offered one of these gongs which I had hesitated to take because I have always felt that, although such honours may be right enough for performers, they draw a writer's teeth. Only by degrees on that evening at the McClellands' I began to scent the reason for my being there. Our hosts had not been told in advance why the Governor-General had asked for me.

Our fellow guests on the occasion were Elizabeth Riddell, a worldly and distinguished journalist, and the British Council representatives, Mac and Isabel Mackenzie-Smith. We were all assembled before the arrival of the personages. Kerr himself made no demands on formality. As a longstanding friend and legal colleague of his host he had every reason to feel at ease.

But one look at her ladyship and the climate perceptibly altered, the temperature fell. It was obvious she would expect protocol observed even in private circumstances. Nancy Robson, the judge's ex-wife, had made it clear to Freda McClelland that Nancy had become Anne, but that her subjects should address her as Her Excellency. One of the most natural and charming of women, Freda seemed surprised to discover such pretensions at what she had planned as a dinner for friends.

Always a bad sign, the Governor-General's lady tended to avoid the women. Isabel Mackenzie-Smith appeared overawed, but I could hear threaded through the conversation Betty Riddell's distinctive laugh. In the dregs of summer, Anne Kerr had got herself up in furs, and gloves to the elbow. She was a tall, stringy figure with what she herself obviously considered allure, and from what one heard, had converted others to her own opinion.

I had more contact with her on subsequent occasions than on this strange evening in the early days of the Kerrs' elevation. The Governor-General struck me as being harmless enough: an amiable, rorty old, farting Falstaff. He continued putting away the drink. Who doesn't at Sydney dinner parties? During the meal Freda's children and some of their friends were brought on briefly. One boy was introduced as the brother of Kate Fitzpatrick the actress. 'Oh,' said the Governor-General, extracting himself for a moment from his wine and the conversation, 'whose little brother – or sister – are you?'

After dinner, as the other guests were moving away, I was pinned against the table to the tune of a few vice-regal farts, and the subject of the Australian Order was broached. 'If you don't take it,' I was told, 'you'll ruin everything.' It was disconcerting, to say the least.

After a day or two I agreed to accept, so as not to ruin everything, and there began a kind of love affair with the Kerrs; private telephone numbers and invitations were exchanged.

I was asked to an intimate dinner at Admiralty House, the only other guest a woman writer whose work and character I had always respected, till her continued loyalty to Annie Taggart/ Nancy Robson/Anne Kerr became difficult to understand as history was made and the dramatis personae came out in their true colours. On this night the dinner, the most succulent imaginable, was served with all pomp and ceremony on a small table beside the fire in the main reception room. What was a delightful occasion at the time became a shameful one later and for ever after.

At the height of the courtship we invited the Kerrs to Martin Road. I can't remember what we ate, but my cooking was not up to that of the chef at Admiralty House. There was plenty of drink, conversation, and the McClellands to help us out with the Kerrs.

Again Lady came in gloves and furs (at least it was winter) and perhaps because she saw the event as a gesture towards demi-bohemia she had streaked her eyelids with green shadow. Discussing the evening afterwards as we cleared away the glasses and cigarette butts, Manoly saw her as an elderly lizard. An intellectual one, it had been borne in on us. Hadn't she been to the Uni? She refused to air her French, in which she is said to be fluent, perhaps thinking it wiser not to run the gauntlet of those who had a bit of the language themselves. But she roamed around, looking at paintings, explaining to herself aloud why her trained university mind should accept this one or reject that. She seemed to see my work-room as a chamber of horrors. 'Oh, no,' she recoiled from one wall in particular, 'that's Cruel Corner! I'm not going to look at that!' The drawing which most offended her sensibility had been done by one of her predecessors, Maie Casey, inspired by the Frances Cornford poem, 'O fat white woman whom nobody loves . . .'

It was the last time we met Lady, the last time we met Sir – Her Majesty's representatives. Towards the end of the

26 First night of *Big Toys*.

27 First night party after *Big Toys*: P.W., Kate Fitzpatrick, and friend.

28   Manoly, Greece, 1978.
29   Lizzie the Carnoustie lass in her nineties.

30   P.W. under the pergola at Martin Road, 1979. (Photograph taken by great-nephew Edward Bishop while aged 8.)

evening the amiable farting Falstaff seemed to be trying to winkle out of us what we really thought of Whitlam. We had nothing to subtract from our opinion of the man we had elected. If Kerr gave the impression of being inwardly confused we put it down to the drink. They were driven away, and we did not seriously question his integrity till the coup.

It was again the evening of a dinner party, this time in honour of David Campbell the poet, who had won the award I set up with the money from my Nobel Prize. I was drudging away in the kitchen about five o'clock when I switched on the radio hoping for distraction from the boredom brought on by chopping and stirring. Like a stream of lava, out poured the news of what was happening in Canberra: that the Governor-General had dismissed the Government elected by the Australian people. Too much has been written already about this shoddy episode of our history. I shall not go into it again in what is intended as a self-portrait. I introduce it only because of the effect it had on me then and afterwards: the part played by the Liberal Senators, the conniving character who replaced Whitlam as Prime Minister, and the more sinister intrigues of the media and foreign powers interested in seeing Whitlam brought down, all contributed to the bitterness which had been growing in me, and which helped drive me farther to the left.

Domestically, we were faced with the dinner for David Campbell, food burnt at the edges, and unresolved lumps in the béchamel. The guests began arriving: David and his wife, Brett Whiteley and his, again Elizabeth Riddell, Virginia Osborne – all of us variously stunned and/or anaesthetised. In our disbelief we kept the radio running throughout the meal, over and over the same sound-track, hoping that something, some slight flicker of hope, might awaken us from a nightmare. But nothing did, and Brett Whiteley passed out on a divan in the middle of the room, which we keep for that purpose, but which nobody had made use of till then.

The Australian community was split on 11th November 1975 and has remained so. The childlike mentality of so many Australians was easily terrified by British, American, and Liberal-controlled media propaganda into thinking they had escaped losing their all through reforms the Whitlam Government was introducing. This supposedly sophisticated country is still, alas, a colonial sheep-run. The return of Fraser and the Liberals at the election of December 1975 plunged many of us into the depths of despair.

Persuaded by Kerr to accept the Order of Australia because I 'would ruin everything' if I didn't, I now felt bound to return the wretched object, and did. To keep it would have seemed like taking a bribe. I have to mention this here because I sometimes detect in others suspicion that I did it in a moment of malicious caprice.

From overplaying their parts in the tragedy for which they had been cast the Macbeths of Yarralumla were reduced to figures of farce. They continued dragging to official functions, reviled, booed, pelted, briefly accorded a kind of martyrdom by those who had set them up till it was seen what a liability the man was. The press showed us Sir lying on the ground at a country agricultural show, apparently shrugged off by the shoulder of a prize-winning beast, while Lady stood aghast at the rump of the Cow who Missed her Opportunity. There were other photographs of Sir, pathetically anachronistic in his top hat and Edwardian clothes pasted on a contemporary Australian background, and the classic shot of him in Ascot grey staggering towards the camera to present the Melbourne Cup.

A peaceful morning towards the end, I was standing on the kerb of a Sydney suburban street when Lady was driven past in one of the vice-regal limousines. She was sitting alone on the back seat under a large flat mushroom hat, holding to her flat stomach a large flat handbag as though it were a hot-water bottle. She looked neither to left nor right.

It is not surprising that Kerr the puppet was thrown aside eventually by Malcolm Fraser, the man who used him in his own rise to power. Or was it Fraser? And will Fraser too, be thrown aside like a flabby puppet by those greater powers who needed him to achieve their purpose?

One hears of Sir and Lady living with their perks and diminished fame in a house in stockbroker country outside London: John Kerr, the Balmain boiler-maker's brilliant son, and Annie Taggart who as a student was innocent enough to face an audience with her fly undone while playing a male part in a university production. I am inclined to pity them both till I remember their pretensions, and the occasion when Kerr as barrister lost a case for one of my friends. When my friend protested, 'But it isn't just!' his advocate replied, 'You should be old enough to know you can't expect justice of the Law.'

Towards the end of 1979 somebody asked one of Kerr's daughters how her father occupied himself. 'Oh,' she said, 'he's taken up cleaning and cooking. The woman he married doesn't care for that sort of thing.'

Marrying the daughter's remark to her father's conception of justice I sometimes indulge in a fantasy in which Sir, armed with a feather duster, totters in a frilly apron round a neo-Tudor mansion till the smell of burning diverts him to the kitchen, while Lady sits behind those green eyelids, spurring him on in her university French. If there is a Hell, it is probably a trail of broken glass, the humiliations public and private which the feet of the damned are condemned to tread, in perpetuity.

Postscript. In 1981 the Kerrs returned on a visit to Australia where it seems that those who used him in their rise to power are now at pains to rehabilitate him. Kerr may even be canonised. One hears that he aspires to the House of Lords.

4 *The Nolans*

We first met the Nolans while we were staying at Fort Lauderdale, Florida, with Manoly's American sister, Anna. Fort Lauderdale is for both of us a claustrophobic event revolving round Anna's house in a mangrove swamp, on a road leading from nowhere to nowhere, in other words the rest of Florida. Anna, a nurse who lives on cigarettes and coffee, with Grace Metallious on her bookshelf, a TV set flickering silently, and a husband an ex-marine (he died before our second visit), has succeeded in eradicating totally the Greek origins which might have saved her. She subscribes to those origins, but in the jargon of the media and the clacketty-clack of standard Southern States American, which effectively poison anybody's European roots.

During our visits Manoly lives under Anna's roof, and because there isn't room for me as well, I am farmed out unprotesting to neighbours. On this first occasion I had a room with all mod cons and every sign of luxury, except that the signs were ephemeral, the luxury could have been peeled off like a skin, or the veneer it was; the wall-to-wall was criss-crossed by the spoor of insects, littered with corpses of the insects themselves, if they had not succeeded in holing up in corners or along the skirting board, to await in their cocoons the next stage of their insect evolution. A larger, temporarily torpid insect, I was conscious of a cotton-wool air pressing against my lath-wire-and-plaster cocoon as I listened to the liquid calls of invisible birds and the thud of the odd avocado. By the perfection of their unctuous flesh the Fort Lauderdale avocados almost atone for the awfulness of Florida.

Into this awfulness broke the Nolans, Sidney, Cynthia, and Cynthia's daughter Jinx, still a little girl lumped in the back of the station wagon, engrossed in *Studs Lonigan*. We had been brought together by correspondence, our common nationality, and the Nolans' enthusiasm for *Voss*. That the meeting should have taken place in Florida made the situation as unreal as that unreal state. Standing beside the station-wagon in which the Nolans had been driving round the United States, we made what must have sounded extraordinary conversation. I was bemused, suffocated by the cotton-wool atmosphere, Manoly more collected, Anna babbling, Sid flickering on and off in a kind of ectoplasmic code, the schoolgirl Jinx still part of Studs Lonigan's world after being forced unwillingly out of the car. Only Cynthia introduced clarity by telling us that everyone – women, men, dogs – fell in love with Sid. Some of the women had been foolish enough to massage his shoulders at parties, not realising that he hated to be *touched*. After that we started off on a drive through the mangrove swamps, past the gas stations, hamburger joints, concrete wigwams which line the long straight roads radiating for some undisclosed purpose from Fort Laud. It was an unpromising beginning to a relationship of great importance. Taken aback by Cynthia's initial revelations, we got things straight. She became one of my closest friends, one of the women I have admired most; her death was a shock I continue to experience.

We did nothing this first day beyond discover that we spoke the same language and liked one another. We drove and talked. Sid was certainly a charmer fit to bring on any masseuse at a party. We talked. We drove. We ate some not so bad Florida stereotype food in a restaurant between waterways. We walked through a park of exotic trees, all armpit and testes, the exoticism somewhat spoilt by a cute mini-railway. We talked. We talked. The Nolans talked about *Voss*, while I mumbled back, foolishly grateful for their appreciation. At that period, mention of the book which had been wrung out of me in sweat and blood like Rose Portion's Mercy did not make me recoil. It was only later when the gushers got to work, and I was idiot enough to let the book fall into undesirable hands for the making of a film which hasn't yet been made, that I began to experience revulsion.

After a couple of these drives through the Florida flatlands we parted company, the Nolans driving north to New York, Manoly and I returning to Australia from San Francisco. We didn't realise how ill Cynthia was. In New York she went into hospital for what turned out to be a serious bout with tuberculosis.

We corresponded. In *Open Negative*, one of the best examples of her writing, Cynthia describes her experiences in the TB ward of the New York hospital where her illness took her. In the same book she writes about the family drive round the States. She was never at her best writing about Sid, an elusive character for me too. She throws more light on him and their relationship in one of the letters she wrote me at this time, 'Siddy comes to the hospital keening like an old Irish peasant woman, her head shrouded in a shawl.'

Over the years I continued corresponding with Cynthia. Her letters were in a practically illegible hand. She complained that typing strained her back, which one had to accept; she sometimes travelled lying down on the tray of their station-wagon, and on one occasion pain forced her to lie on the floor in a box at Covent Garden during a performance of *Don Giovanni*.

Cynthia was the one who wrote, erratic, extravagant, sometimes crazy letters which would open of a sudden into passages of great sense. Sid always promised to write, always with tears in his eyes, at the last moment at airports. He never did, or no more than a few flickers of code on the back of a drawing. Because he functioned visually I accepted the situation, while resenting the broken promises. I tried to persuade myself they were a side-effect of Irish charm.

What did surprise me was that after our meetings our conversation seemed to have evaporated. It was never like this with Cynthia: conversation always brought us closer; I can still remember snatches of hers. I sometimes wondered how Sid performed conversationally with friends like Robert Lowell and Kenneth Clark. Perhaps he needed a more intellectual foil. I am not an intellectual. There is nothing cerebral about me; if I have something to give it is through the senses and my intuition. For that matter I have never thought of Sid as an intellectual, though he is inclined to throw off those desperate semi-coded messages, or run up a flag with 'Kierkegaard' on it during the soup.

In the Nolan partnership Cynthia was steel to Sid's elastic. She had to guard a husband and an artist against the predators and the masseuses. If it had not been for Cynthia I doubt Nolan would have reached the heights he did in his best period. He would have drowned much sooner in the sea of flattery which sucks so many artists of importance under.

Naturally Cynthia was much hated by those who had designs on Sid. They hated her for her patrician, increasingly ravaged face, her unfailing taste which showed up their own flyblown vulgarity, their ambition to nail down an investment, to bed a genius, or simply booze with a mate.

Cynthia held out for as long as her physical strength allowed. We couldn't have faulted Sid any time we were all together. He was the most attentive husband, even at C.'s more capricious moments, and the most thoughtful friend, fetching us from airports and hotels, driving us to theatres, gardens, or the house at Putney which was Cynthia's masterwork and swansong, the expression of her taste, her complement to the best of Nolan's painting. Physically the house was not unlike Cynthia herself, slender and muted, in it a force preparing to uncoil, or breaking gently — a shimmer of Launceston Tasmania in London beside the Thames. Cynthia professed to hate Australia, but like most expatriates never escaped.

Sid claims to love Australia and returns here for brief spells. His attachment is material as much as spiritual. As I see it, he needs a mother more than a wife and Australia is the great maternal bub on which he sucks. There is a story he tells of himself as a large child sitting behind the driver of a Melbourne tram with his mother still giving him her breast, when the driver turns round and says, 'If you don't stop feeding that kid I'll get off the tram.' The breast, I think, has been his sustenance and his downfall.

My relationship with the Nolans ended, wrongly or rightly, the night Sid rang from London to tell me of Cynthia's suicide. Our previous summer together, the four of us strolling through those miraculous gardens which continue to flower on the outskirts of a decaying London, going to the theatre, listening to music, eating good meals, had been perfect in a way which foretells disaster. By degrees I sorted out the details. After lunching together, and without dropping a clue, Cynthia left Sid, ostensibly to do some shopping, and went to a room she had taken in deepest Betjeman country, the Regent Palace, where Alice from Bootle had rootled with a safety-pin, and where, as a steamy, masturbating teen-ager, I had spent a lonely night with incompatible guardians. Here, in this cheap marble mausoleum, so remote from everything she loved and stood for, Cynthia swallowed the overdose.

I have never been able to blame Sid, knowing they were both, that we are all, always, to blame. If I have not been able to accept him since Cynthia's death, it is from knowing the Cynthia in myself, and that I might have acted in the same way. What I cannot forgive is his flinging himself on another woman's breast when the ashes were scarcely cold, the chase after recognition by one who did not need it, the cameras, the public birthdays, the political hanky-panky, 'if I have to shake hands with Fraser I'll wear a glove' (one wonders whether he wore it the second time round), all of which, and the Athenaeum Club, would contribute to the death of any painter, as he must know from his own recent remark on having a gallery named after him, 'I feel as if I should be dead . . .'

The best of him will never die, nor will Cynthia, because the best of Nolan was their collaboration.

5 My Other Diva

My encounters with Melba are childhood memories, the half-a-dozen words we exchanged by chance in the class-room when she came to my school, otherwise the clockwork lark I listened to respectfully as she sang on the gramophone at 'Lulworth'. My other diva I first heard on the kitchen radio at Castle Hill singing the inevitable arias chosen by suburban girls competing in Mobil Quest: that recurring *Bell Song* for instance. I didn't see Sutherland in the flesh till as an old man I listened to her, a polished and acclaimed diva, in the Sydney Opera House.

I met her finally at a supper arranged by our friends Desmond Digby and James Allison in what I persist in calling anachronistically their 'flat' after a performance of *Suor Angelica*. To me this is one of Puccini's most persuasive operas, scornfully dismissed by some of Australia's 'intellectual' critics, their rational minds offended by that tinsel apotheosis which, as I see it, conveys the essence of Italian Catholicism. Released from the bel canto trapeze Sutherland accounted for the nun in heartrending human terms. I'm pretty sure I'm not a Protestant hovering on the brink of conversion; I have met too many virulent converts. I don't think I am a sentimentalist, though when I was a child, Lilian Gish in crushed organdie had me bursting into ready tears. It was the humanity of Sutherland's interpretation which moved me that night at *Suor Angelica*.

We collected ourselves and drove to the party at Kirribilli. The hosts had cooked themselves to a standstill, the diva was understandably late, myself nervous at the prospect of a meeting with a great singer.

She arrived, with an entourage of husband, majordomo, and Moffat Oxenbould and Graeme Ewer of the Australian Opera. I've been told since that she had not been warned to expect a novelist. That may have accounted for some of it. We were both on edge, myself dressed up far too much in my professorial suit, she avoiding, this large woman, or soft sculpture, in a jujube shift. She gave an Ocker performance from the word go. Melba before her, however immaculate the silvery voice, was at heart an Ocker, kicking a metaphorical football, or the tenor's arse. Understandably, Sutherland tucked into the supper, and while forking it into her mouth assured me from a less occupied corner of it that she had not read a word I had written, but could not put *The Thornbirds* down. Her husband and the majordomo tried to make good the omissions by dragging in *Voss*. If they had but known . . . Anyway, it passed off. I can't remember who escaped first. During the night, when you remember the remarks you should have made, I realised I should have implied that Beverly Sills is my favourite operatic soprano.

The moral of it all is an old one: divas should never meet; though I believe that if I had met Callas we should have understood each other, and that it might have been a relationship such as I had with Cynthia Nolan.

6 D.

I think of the telephone as a lifeline. My mother was an addict. On moving to 'Lulworth' in 1915 I remember her rushing at this still novel machine, winding the handle, and exclaiming, 'It works!' It must have generated equal excitement in her three-year-old child. I too, became addicted, and overcame a little of my shyness in conversations where my vis-à-vis remained unseen. But it was not till living in London in the 'Thirties that I became really hooked. Looking back, I must have lived half my life at the end of a flex, one so long I could carry the instrument into the bathroom to indulge. This state of decadence ended abruptly when I read about a showgirl electrocuted while phoning from her bath.

For Manoly on the other hand the telephone is the great disrupter, partly because in his boyhood there was no telephone in the house in Athens and the children were dispatched with written messages to their aunts' friends. So he did not grow up in the telephone tradition, and always eyes me disapprovingly when passing through the kitchen where we have our phone, as I sit extracting the juice of life from my various purveyors, usually between washing up and sitting down at my desk. He cannot believe in my elixir any more than what he sees as my circus religion, his own faith deriving from the Pantocrator and the sap of plants.

But he accepts. For many years he has accepted D., the telephone's recurring voice. His own exchanges with the voice are limited to 'Yes' and 'No', baring his teeth in the smile he keeps for invisible disrupters (usually people he enjoys meeting face to face) before handing me what looks like a hot receiver. He leaves the kitchen.

D.'s openings are usually muted, gulped back into the hole at the end of a funnel.

. . . are you a ray of sunshine today?

. . . couldn't expect me to scintillate – not as things are . . .

. . . No-ooh . . .

. . . haven't heard from you lately. Your grazier friends must have been up . . .

. . . Arr, *look*, if you're going to carry on like that . . .

. . . If you're still under their spell I can't very well help . . .

. . . You don't know what you do to people . . .

. . . apparently never enough . . .

. . . I'm going . . .

. . . No need to get hysterical.

. . . I'm not. That's over. I haven't taken one in months . . .

. . . What do your graziers have to say for Fraser by now?

. . . They're on with Hawke.

. . . That oughter show yer.

. . . There! D'you hear that noise? It only ever happens when I'm talking to *you*. We're being bugged.

. . . What if we are? Let's give 'em something real juicy . . .

. . . But it hurts my ear . . .

. . . Buggery hurts. Remove the receiver for a few seconds and start again.

. . . Last time it happened my ear was sore for the rest of the day.

. . . Why isn't mine?

. . . Arr, look, I'm not in the mood, I'm going . . .

. . . Wait a bit – something I forgot to tell – something *awful* that's happened . . .

. . . In *your* book only the awful happens . . .

. . . But if it does you can't ignore it.

. . . wrecked my morning.

. . . brought up soft — not enough slaps — and too much sussy butter.*

D. roars. We always enjoy a good roar together.

. . . Poor old Nanna's sussy butter. She died of cancer . . .

. . . at a fair age.

. . . Only sixty-seven.

. . . I'm rising sixty-eight.

. . . You are — ha-ha! . . .

. . . If it comes it comes. Everybody's got it.

. . . M. went to the Philippines before she died. F.'s got it in the bowel . . .

. . . deal with it in the bowel . . .

. . . not all that easy if you're not a Christian Scientist . . .

. . . Half Australia is. They wouldn't have got themselves Fraser if they weren't. A lot of kidult Christian Scientists . . .

. . . I've gotter go — gotter look at my chicken . . .

. . . You were going to tell me what to do with New Zealand whitebait.

. . . Well — you thaw it — leave it in its liquor. Then you . . . Look, I'll write it out — Mummy's own whitebait dollops. But I gotter go . . .

. . . Bye then.

. . . Bye . . .

. . . But you gotter admit it's the Age of the Crab.

. . . Any old age was that. They say they're worst when you get 'em in the armpits . . .

. . . or lice in the head. It used to be only the abos — now they have it in the nicest schools . . .

. . . I'll tell you something — hepatitis and herpes are rampant down in Melbourne — epidemic proportions . . '

. . . It's those tight jeans . . .

. . . or somethink . . .

. . . that's right — or somethink.

. . . It's the times. You gotter face the times . . .

. . . You gotter. Bye then . . .

. . . Yair – for now . . .

I give D. hell, but D. enjoys it, as do most of those others admitted to hell – or so I dare believe. I also believe tunes played on the tin whistle have their own importance. Its themes give a meandering lead through the dense thicket surrounding them, over and above the melted-toffee surge of strings, in and out the convinced but not always convincing affirmations of brass and drums. I need to play my tin whistle and be answered back.

The telephone is ringing. I'm on my way. It's probably D. The kitchen is littered with last night's pots and dishes, thick with last night's fug. The telephone . . .

. . . Yurss?

. . . *Well* . . .

. . . wondered what had become of yez . . .

. . . Look – we've been going through it and you never bother to ask . . .

. . . just what we say – sitting there thinking only of themselves . . .

. . . if you knew . . .

. . . I like to think we . . .

. . . I'll ring off if it's like that . . .

. . . in a glass tower. Trotting across to the Opera House – picking up *her* petty point when she drops it. As for the Yellow Canary . . .

. . . You don't know what the Canary's up against – got to unravel – please them all – once in power you're a politician . . .

. . . I'll buy that – not all of it . . .

. . . It's true, though . . .

. . . What's true I – wouldn' – know . . .

But keep on playing my tin whistle, in hopes. It could get you there quicker than Mahler's moans and Bruckner's brouhaha. Thank you, D., for joining me on tin whistle; a duet is more consoling than solo.

7 *Jimmy Sharman and his Acting Troupe*

My Uncle Clem Withycombe used to say, 'If a person doesn't collect a few young friends, he'll find he's the only one who isn't in the cemetery.' Uncle C. used the word 'person' long before it was in fashion. He was a non-believer, but gave to all the Churches on the off chance. He would lay a bet, and often lose. I would go to him, a countryman with a vast store of country knowledge, when I had to check up on the construction of a wagonette or how to drench a wormy sheep. He was also a shocking gossip. This too, is invaluable to a novelist. All of which is of course a digression. Uncle Clem died years ago, and true to his contemporary leanings, did not end up in the cemetery but in the Northern Suburbs Crematorium, where an anonymous parson got his biographical details wrong. Though I mourned and missed him, I didn't think much more about his advice on collecting younger friends, and anyway he was wrong there to some extent: a person doesn't collect, he is collected. As our older friends began going to the cemetery, and one no longer saw eye to eye with some of those who remained, I was wrong and vain enough to think I could get along under my own steam — just as years before the egotist in me had rejected God as unnecessary.

As happened when I fell on my back in the mud at Castle Hill and started cursing a God who did not exist I was led blindfold one night to the still rather crummy Jane Street Theatre and a topical revue *Terror Australis*, which a respected critic had panned. We were a handful of an audience. Director, writers, actors, were all young. As they exploded at us they had nothing to lose but success. I was both exhilarated and frightened, so much so that I continued feeling exhilarated during the long trek down to the main road where I finally succeeded in trapping a taxi. My enthusiasm did not wane overnight, and next morning I wrote the paper a letter questioning their critic's verdict. It did not save the revue.

So much for that. Some months later I was coming out of a city shop after buying a gramophone record. I heard an urgent padding behind me, and an even more urgent voice, 'Hi, Mr

White . . . ' It was a scruffy young man in T-shirt, jeans, and sneakers. He wanted to thank me for the letter I'd written to the paper praising his condemned revue. We stood together in the street, made awkward by our confrontation, our different worlds, the gap between our ages, in spite of what we had in common in that abrasive, shit-slinging revue. We drew apart. Though gratified, I did not imagine I'd have any further dealings with Jim Sharman.

He was the wonderkid, director of *Hair* and *Jesus Christ Superstar*. As a child he toured the country agricultural shows with his father's boxing troupe. Though Sharman the Father was a Roman Catholic, his wife was a Prot. Jim the Son once told me after we became friends, how much he regretted missing out on a Catholic childhood, but had got into theatre instead. I believe that most Australian artists of Protestant upbringing would admit, most of them grudgingly, that they had missed something by not experiencing a Catholic childhood. It has always been obvious to this outsider that those who have left the Church luxuriate even in their lapse.

As a child I preferred the tentative theatre of matins at St James's, King Street, to those bleaker services at Bong Bong (where at least we had the blue riband above the altar flaunting its questionable message GOD IS LOVE in letters of gold) or in the asbestos church at Mount Wilson, filled with gloom by the tree-ferns which guarded it too closely, musty air from its month's closure, and the disturbing wheeze of the harmonium.

My mother introduced me to actual theatre at an early age; we were always going, and I cannot be grateful enough for that. It filled some of the gap in what was by normal standards my solitary childhood and youth. If I had friends they never entered, or more truthfully, were not allowed to share that mysterious private world of a child meant to become some kind of artist. My vocation came closest to revealing itself in those visits to the theatre, usually musical comedy, in the early bubblings of sexuality, and expeditions through the streets observing, always observing. I suppose I was happiest visiting elderly literate women, book shops in which the smell of books, the feel of them, the titles I read, intoxicated, and most of all, during the hours I spent at 'the pitchers', either at the Cross or

244

down George Street. My mother did not approve of the cinema; films, she maintained, hurt her eyes, and picture-theatres she considered common. But she did not object to my going to the pictures and was not unduly censorious of what I saw. Perhaps because she could not take the cinema seriously she did not think her peculiar child could come to serious harm.

Decades later, Jim Sharman described for me his own solitary youth, roaming as a schoolboy round Coogee and Bondi as I had roamed Darlinghurst and George Street, mooning, dreaming, going to the pictures. Discovering this bond partly explains our relationship.

I didn't get to know Jim till after a letter from Ken Southgate of the Old Tote Theatre asking whether I would consider letting Sharman direct my play *The Season at Sarsaparilla*. Three of my plays, including *The Season*, had been staged in Adelaide years before, and a fourth in Melbourne. There had been a handful of enthusiasts. Otherwise all four were dismissed by critics and public, and ended up on the shelf. I was amazed that the wonderkid of Australian theatre, idol of the young, should want to disturb the dust. But it didn't seem so unlikely when I thought it over; in fact it occurred to me that Sharman might be just the one to resurrect *The Season* in the 'Seventies. It surprised me I hadn't thought of it myself.

He came to see me, pink and white above a royal-blue windcheater. We sat together in my work-room, rather over-awed by each other, while exploring the possibilities of a theatrical collaboration. What gave him confidence, I think, was that I could see the play in terms of his generation, that it did not have to be cast with actors from mine. During the talk there was a thread of I don't know what (a spider's?) hanging from one of the pink-and-white cheeks, visible when air lifted and light caught it. I have never told him about this thread, or how I wanted to lean forward and sweep it away. In one way or another I might have ruined everything. We parted company without any embarrassment, and the production started taking shape.

As soon as the news broke, some of my contemporaries began exchanging smiles which implied, 'The silly old fool is asking for it; he'll come a cropper. Poor Manoly!' None of it happened.

Manoly has always liked Jim – as far as you can know what those closest to you really think and feel. On my side it would be dishonest if I denied an occasional twinge of sexual jealousy in my relationship with Jim, but our collaboration settled into what I would call a perfect example of fruitful sublimation.

Sharman's revival of *The Season at Sarsaparilla* became the success it had not been in Adelaide, Sydney, and Melbourne fifteen years earlier. In saying this I don't mean to condemn the original production. I remember its vitality and the care devoted to it, but I also remember sordid squabbles with a director whose temperament was too close to my own, and the reactions of audiences incensed at being faced with their own suburban lives. By 1976 the public and some of the critics were able to grow sentimental about humble beginnings. They could accept the comic side. On the first night, in one of my more ridiculous fits, again wrongly dressed, in what has been referred to as my 'mafioso suit', I made a speech from the stage of the Opera House Drama Theatre to an establishment audience whose lips I could sense curling like petals in a westerly. It didn't matter. I had to thank the director and the actors, some of whom are still my friends.

Voices persuade me to write plays. As a result of the revival of *The Season* I wrote *Big Toys* for Kate Fitzpatrick and Max Cullen who had appeared in the earlier piece. *Big Toys* poured out, just as I had dashed off *The Season at Sarsaparilla*, the first draft at Castle Hill, the second in the writing-room of a coastal steamer on a voyage between Sydney and Brisbane, bound for Fraser Island which became the backdrop for important scenes in *The Eye of the Storm* and *A Fringe of Leaves*. Looking back, the overlapping is what surprises even myself: that there should be any connection between the two novels and a theatrical folk piece like *The Season*. Voices are understandable as the progenitors of plays: the voice of the Perth actress Nita Pannell, who appeared in some of mine, but not all those she influenced, and later the voices of Kate and Max.

Directed by Jim, and designed by Brian Thomson, another Australian theatre genius, *Big Toys* upset the vulgarian shyster rich as *The Season* outraged unsophisticated surburbia years before. What I set out to show in *Big Toys* was the corrupt state of Sydney society today. Members of this society who did their cultural and social duty by coming to the play resented its portraiture and politics. Fashionable ladies were most vehement in their rejection, though the more honest among them admitted it was too near the bone. Someone visiting an exhibition of paintings by an artist considered a sound investment came across a ring of Mag Bosanquets deciding which of their friends was the original of Mag. On the other hand I was told I knew nothing about the Law: Ritchie Bosanquet was the caricature of a barrister, and a spiteful one at that. Then we began hearing he was not *unlike*, that he *was* this one or that other to the life. But *Big Toys* remained unacceptable: too black, too suicidal. Time has rolled on since then. The play was filmed for television. It made no great impression on those who watch in the glass towers and the lounge-rooms of their homes. Perhaps we need another fifteen years.

After the collapse of the Old Tote, Jim Sharman and his associates persisted in their attempts to contribute to living theatre in Sydney. There was the ill-fated venture at the Paris during which sensibilities and bank accounts were somewhat damaged. Jim must have suffered intolerably, but as usual there was scarcely a visible sign as he slipped in and out the dark weeds of his personal aquarium. He grew paler certainly, and hunched. Actors say he is cold, not an actors' director. To me he is for the writer. He respects one's text and one's feelings. Visually, particularly in collaboration with Brian Thomson, he is superb.

At the Paris he had planned for Dorothy Hewitt's *Pandora's Cross*, Louis Nowra's *Visions*, and if the money held out, White's *Cheery Soul*. In spite of effective moments of theatre and Thomson's stunning set, Dorothy's play failed, and an opposition was formed. It was the depths of winter, the theatre underheated. A frozen and perverse public condemned Rex Cramphorne's production of Nowra's brilliant play. The acting of Kate Fitzpatrick and John Gaden, and our first glimpse of the emerging talent of Judy Davies, could not save it. The Paris season floundered. It seemed unlikely that *A Cheery Soul* would ever be seen again.

Jim has said since, 'If you stick around long enough it doesn't rain but it pours.' We stuck around, and if it didn't pour, *A Cheery Soul* was staged at the Opera House, the first production in an attempt to replace the Old Tote with a State Theatre. We broke the box office record at the Drama Theatre. Jim and his troupe, with Robyn Nevin as Miss Docker the Cheery Soul, and Brian Thomson's jack-in-the-box sets had Sydney wondering more than ever what it ought to think about an unorthodox proposition.

Since then there has been the film of *The Night the Prowler*, a success with a few, but a failure with that great Australian majority which prefers rose-tinted soap opera, and Sharman's peak so far, the Britten *Death in Venice*, again a visual miracle, in collaboration with Thomson and Luciana Arrighi. Just before the opera opened at the Adelaide Festival Jim was appointed the Festival's next artistic adviser. What this will bring forth beyond some kind of unexpected excitement, one cannot foresee; one only knows for certain that 'If you stick around . . .'

I've stuck around. If I dare come out and say I feel as though these younger people are my children I shall probably be laughed at, accused of harbouring delusions, and of hoping to revive spent powers by associating with youth. I don't believe it is altogether that. We shall always have something to give each other, even though age ensures that my contribution to these relationships can only be indirect and passive.

In May '79 Sharman staged a production no less memorable for being minor. It was my birthday. We were summoned to the house at Enmore, to what we thought was going to be dinner with the two Jims. We arrived and walked through a dark silence, but which led into an illuminated cave, at either end clusters of these younger friends, or elective children. A shock for two old men who had come expecting a quiet dinner. We blenched, but got ourselves together as the ceremony exploded round us: the tinsel, the table spread with Jim the Younger's food-offerings, reflections in mirrors and the tropical pool in the jungle beyond, the champagne necks, the enamel butterflies, the drorings – Kate, Martin, Elisabeth, Kerry, Luci, Frank, Fran, Linda, the Coleings – the lot. If there was bitching too, that was part of the party.

Luci drove us home, her children asleep in the back of the comby, still signs of life in the side streets, sinister in some cases. Outside a service station cars had been set on fire. They were blazing away on their own. One felt these smoking torches might be lighting the way to some greater bonfire. For the time being, comforting delusions, dew dripping from eaves and leaves, mopokes in the holmoaks, waterfowl hooting from the lakes in the park, Luci's good-night kisses, the Australian sleep . . .

8 *The Nobel Prize*

There had been warnings. A bunch of white roses was brought to the door that morning. But nothing official. As we went to bed we had no idea of the awfulness awaiting us. I had dozed off, but was jerked awake by a bell ringing, followed by a banging on the front door. It had happened all right. Never the best watchdog, Manoly went down to face whatever we had to face. An advance phalanx of reporters told him I had won the Prize and that I had better come down. He said I had gone to bed and that he didn't think I would see them at that hour; they should come back in the morning, I was always about by six. They shouted at him that I wouldn't get international publicity if I didn't come down at once. He said it wouldn't worry me, and repeated that I'd see them in the morning as early as they liked. He closed the door.

The battering began, front and back, dogs barking from laundry and kennel as the representatives of the media tramped around outside the house. Shouting. A trendy female reporter roused a neighbour with whom she was acquainted, to plead with me to expose myself in the middle of the night. Shouting, tramping continued. A bunch of them camped on the front lawn with their lights and cameras. They only went away on realising my stubbornness surpassed theirs.

I kept my word, went out at the promised hour, and received the returning media. I sat on the front veranda, or was dragged out on the lawn by those who wanted full light. I was no longer a human being. I was the object their profession demanded. Questions were asked interminably. How ridiculous many of these were did not matter in the least; the pervs and parasites had to get on with their job. I sat all day answering questions, facing cameras, till seven o'clock in the evening, when our visitors started tailing off. An emotional Finnish lady panted up the path, afraid she might have missed out on the interview for which she had been flown to the other end of the world. A cool young woman from Singapore, with all the practicality of the Chinese, had never missed out on a thing. I would have liked to show her that this was the one occasion, but by then I was too listless, or relieved to have got it over. I forgot that I still had to face the academic vultures. Manoly who in some circumstances has more presentiment than I, predicted, 'Our lives will never be the same.' He was right.

I have mentioned going to a dance every night for a fortnight when a youth of seventeen, to please my mother. At the end of that unholy fortnight I never went to another. My reactions after the Nobel Prize were much the same. I had to satisfy those who had awarded it by accepting the grotesque aftermath. But surely it is understandable if I have scarcely given an interview since? What I could not accept at the time was the invitation to fly to Stockholm and receive the award in person. This refusal must remain incomprehensible to all those who don't understand my nature or my books.

It was suggested that Sid Nolan represent me. It seemed to me a sensible solution: the Nolans were at least in the right hemisphere, and Sid delights in pomp and ceremony as much as I recoil from it. The Nolans went. They had trouble tying the dress tie for the great occasion, and how I sympathised, remembering struggles of my own, referring with little result to a diagram as I stood in front of the glass. They enjoyed some of the fringe events, but the night of the ceremony induced such fright in Sid he told me he took to drink and didn't get off it till long afterwards. I have felt guilty as a result. Am I possibly responsible for the break-up of an artist, the relationship between a husband and wife, and a close friendship, by refusing to face the music?

Certainly, as Manoly predicted, life has not been the same since.

The Nobel Prize may be all right for scientists. It was all right for writers in the days before communications developed. Read Willy Yeats's account of his dignified progress to Stockholm by sea to receive 'the Nobel Bounty'. His father, J. B., much more a figure of the present day, would understand my writing so ungraciously of what is intended as the supreme literary honour.

9 *What is Left?*

Memories – friendship – love, however thin its ice – food, if teeth allow – sleep – the dark . . . Would it have been any different had I begotten actual children? I doubt it. I'd have been a rotten, irritable parent. There would have been all the misery of disagreement. I doubt whether I could have kidded myself, like so many parents, that my continuity was assured.

Possessions are valueless. I have never hung on to heirlooms. Those which remain are my father's concave, ivory-backed, gold-monogrammed hairbrushes, and Clem Withycombe's carriage clock. The discoloured bristles of the brushes are teased from time to time with a pin, and give up the fuzz of an heir's dingy, fallen hair. The clock clangs wrong quarters of the hour. So much for the heirlooms. I can't divest myself of memories; they cling like cobwebs, while blood is the river which cannot be crossed.

You reach a point where you have had everything, and everything amounts to nothing. Only love redeems. I don't mean love in the Christian sense. To lavish what is seen as Christian love, indiscriminately on all mankind, is in the end as ineffectual and destructive as violence and hatred. Love in homeopathic doses can be more effective than indiscriminate slugs of the other doled out to a sick society. Christian love has lost its virtue, as antibiotics lose theirs through over-dosage.

Christians will say I don't understand Christian love. Perhaps I don't; it is too grand a theory. When I say love redeems I mean the love shared with an individual, not necessarily sexual, seductive though sexuality may be. Those who believe I don't understand Christian love will probably be joined by the ones who interpret this other statement as the straw grabbed at by an ageing man as passion floats out of reach. If it is making do, let us make do, whatever our age, in a world falling apart.

I read constantly that I am a misogynist. If I had married a woman or two and jumped into bed with three hundred others, treating all of them abominably, I'd have been known as a woman lover. Only the feminists, understandably, would have accused. In life I have known far more admirable women than admirable men. Those who have read my novels attentively, not just glanced through one or two of the more controversial at the time when they were conversational fodder for dinner parties, must surely have seen this. Of course my women are *flawed* because they are also human beings, as I am, which is why I'm writing this book. Every day as I sit down at my desk I struggle to overcome a revulsion for what I am doing. But it had to be done.

Children who used to look up at me have shot up Australian-style into blind-eyed gods and goddesses, and now look down. One such young god whose mother recently asked me to 'help develop his imagination' I meet in the street. The mother had started telling me about the books he must study this year in school, 'something in French – by Camus – I can't remember the title – and the English books I can't remember, but ask him.' He is one of those children who in the mother's opinion, 'would like to be artists if they didn't have to earn a living.' So on this bright morning of early autumn, the ground mist in the park not yet dispersed, I start on one of her would-be artists, the young god advancing towards me down the pavement.

ME: . . . the Camus book your mother told me –I expect it's *L'Étranger*.

Y.G.: Yair, that's the one we're doing now . . .

He is wiping the sweat from his cheek. Is it such a terrible experience to be asked what was only half a question? I try again.

ME: And what about the English books?

Y.G.:   (grunts, almost groans at this old man's perversity) . . . can't remember . . . (gasps again, this time brushing the sweat from his forehead) I'll let yer know . . .

I let him go. He stalks carefully out of range. Can he have jogged himself mindless? Or is it simply that I am *l'étranger* to this Australian adolescent, putative artist in his mother's book? If I live long enough to communicate with him at all he will address me in the standard tones of the Australian lawyer or doctor – politician? no, poor boy, he is too shy to stray so far afield. As for the artist, there is no faint flicker that has not been effectively quenched, by upbringing, education, organised sport. The artists are here all right, but they have to elbow their way against the surge of the colonial sheep race.

The smaller children are still a joy, though Sara V. has split into two characters, the pure, scraped-back little girl in her convent-school uniform playing the part the nuns have started rehearsing her for, and Sara later in the evening slommacking up the street in her high-heel drag, pouting, chewing.

SARA V.: (self-consciously, through a mouthful of gum) I've got to do a project on you.

ME: What do you know about me?

SARA V.: (grimacing, blowing a bubble with her gum) I haven't started yet.

Another, smaller Sara, enjoys a world of fantasy, while never far removed from the crushing reality of childhood. She wants to be big and strong, like a man, 'because men *do* things.' Boys admire her for her Red Indian blood (not a drop, but her mother is Canadian.)

This afternoon Sara R. is standing in the thin shade of the mulberry tree, naked except for her long pale hair and her briefs — and the shade and her ruthless innocence. She is holding a glass beer-mug half filled with green, synthetic lemonade. Between us a perforated wall. The flow of traffic along the road collides with the flow of her information, as Kattegat on Skagerrak. So I am denied the half of it. By flashes I get the reason why she is not at school. The symptoms of her diarrhoea. How she got up in the middle of the night and ate a lot of chocolate — 'the worst *thing* . . . ' Sara R. is a born back-fence gossip, a gift to any novelist, perhaps a novelist herself.

SARA R.: . . . had our first fight . . . this boy Andrew, age fourteen . . . Then there's Nicholas . . .

I gather the fight has been between their new next-door neighbours. I tell her I feel sure such a big strong girl must be able to keep a fellow of fourteen in his place. Could she have begun to doubt? She is rather expressionless as she stands holding the mug of green lemonade. But starts again.

SARA R.: . . . Andrew and Nicholas . . . Then there's Georgie and Cassie — the girls are adopted.
ME: I've seen Cassie — she's the red one . . .
SARA R.: Cassie's blonde.
ME: But I haven't seen Georgie.
SARA R.: No, you haven't. But Georgie's seen *you*!

It deflates me a little. I say it's late. I've got to get on. But I tear myself away against my will from this omniscient presence in the shade of the mulberry tree.

As I walk past with the pugs there is the smallest girl, probably a casual, at the flats. She is holding a tabby kitten called Seymour.

'What's the dogs' names?'
'Daisy and Pansy.'
'Is one a boy?'
'They're both girls.'
'The Pansy one looks like a boy.'
True — Pansy has a hooded vulva.

Seymour's owner runs after us. 'What's *your* name?'
'Patrick.'
'What's your wife called?'
'Haven't got one.'
Scarcely credible.
'Any cats?'
'Ugly and Gough. And two more dogs – Nellie and Eureka.'

I walk on, realising too late that I have not asked my friend's name, when names are so important, and she may be so casual she'll disappear before I find out.

As it happened she didn't disappear, and she's another Kate. Since our recent meeting this persistent child has forced her way into our back yard.

Nothing casual about the neighbouring adults; they're here till they're boxed up, for the most part elderly Irish Catholics born in the houses they'll die in. Together we've gone through illness, weather, political crisis, threats of demolition. We no longer bother to hide behind the garbage bins the bottles we put out in the lane. Dogs recognise you by your smell. Lawyers and doctors are wary. I do better with the exotics. Arki finds out for me where I can buy instant *dashi*. Esmé the black grandmother has realised that I too, am black. We talk about gardening and the dangers and advantages of swimming-pools. She agrees that pools can bring on pregnancies. We have warmed up since Esmé's Rufus attempted rape on my spayed Eureka.

Eureka – this big red rumbustious hopeless mongrel dumped in the park by her owners on the eve of her first season. After she had lain three days moaning the other side of the railings we took her in out of sentimentality, and never stop regretting it. Loving her at times. A galumphing galoot if ever there was one. Betjeman's Primula Guest. Who without a thought takes a pug's skull in her jaws, albeit affectionately. Licks the skin off your arms. Tears your pockets by sticking her paws in them. Draws blood with her dew-claws. Bashes shrubs to smithereens with her tail. Devours a bonsai or two. And in the beginning ate shit, not surprisingly after being left to starve. Now her eyes are not so blank. They are the painted eyes of an odalisque. Her vacuum-cleaner of a black muzzle has grown more sensitive. If you gave her the opportunity she'd get into bed with you and chew up your blanket before dawn — Eureka the found one.

As we creak round the slopes of Mount Meroo in days to come, joints locked, vision increasingly blurred, the voyeur may see us as papier mâché versions of monsters left over from the pre-historic landscape — or with the brutality of half-knowledge, that bloody pair of poufs. Whereas if spared thunderbolts and bonfires, I'm sure we'll live on like more orthodox couples who have weathered a long relationship, bumping against each other in the dark, looking for and finding lost objects, playing accord-ing to the moment one of the several characters in our repertoire: husband/wife, father/mother, continuing child, to name the more basic. There will be the same squabbles and resentments, and in atonement, an exchange of small kindnesses.

Early morning has always been the best time of day. In childhood, gold pouring through the slats as I got up to raid the pantry for crystallised cherries, finish the heel-taps on the supper table, and settle down to the plays of Shakespeare. Now when I wake, the naked window is washed pale. As I use the eye-drops the first bird-notes are trickling in. Down in the garden, light is a glare. I'm forced to bow my head whether I like it or not; the early mornings of old age are no setting for spiritual pride. Spiderwebs cling like stocking-masks to faces that blunder into them. Dogs point at vanished cats, follow the trail of the night's possums. At the end of her lead Eureka bays and threatens to pull me over in a cataract of light, scents, dew. We collect ourselves as far as it is ever possible.

If I were to stage the end I would set it on the upper terrace, not the one moment of any morning, but all that I have ever lived, splintering and coalescing, the washed pane of a false dawn, steamy draperies of Sydney summers, blaring hibiscus trumpets as well as their exhausted phalluses, ground mist tugging at the dry grass of the Centennial steppes, brass bands practising against the heat, horses cantering in circles to an accompaniment of shouted commands, liquid calls of hidden birds, a flirt of finches, skittering of wrens, bulbuls plopping round the stone bath carved by Manoly in the early days at Castle Hill, as though in preparation for the Twyborn moment of grace.

Notes

and disciple of D. H. Lawrence. She followed him to New Mexico and continued living there after his death.

136 9 **Akubra:** well-known brand of Australian felt hat.

147 22 **David Moore:** author of works on the Australian Aborigine.

147 23 **John Young:** Professor of Physiology, University of Sydney.

147 23 **Klari Daniel:** a Budapest Jewess who migrated to Australia with her family after persecution of the Jews began.

154 33 **Professor Leonie Kramer:** Australian academic.

160 15 **loukoumi:** Turkish delight.

160 17 **kritharakia:** small pasta.

162 9 **komboloyi:** a version of the rosary which has lost its religious significance, now a plaything to calm the nerves.

172 15 **kendron:** combination of a café and a restaurant, where you can have anything from a cup of coffee to a simple meal.

172 33 **Capo d'Istria:** played a great part in the uprising against the Turks which led to Greek independence. He became President of the Greek Republic in 1827, but was assassinated at Navplion in 1831.

173 17 **tsipoura:** a fish by now practically extinct in the Eastern Mediterranean.

175 21 **Dickerson:** Australian painter.

184 24 **Karagiosis:** a puppet show, Greek version of the Punch-and-Judy.

186 9 **pallikaria:** literally 'valiant men', often applied to the fighters for independence.

188 18 **vaporaki:** steamer.

188 33 **skalopatia:** steep cobbled path with steps set in it at intervals.

203 13 **tsipooro:** plum gin.

203 13 **tsitsiravlo:** pickled shoots of the pistacia.

217 16 **tyropitta:** cheese pies.

241 5 **sussy butter:** sugared butter.